Excel Crash Course for Engineers

Eklas Hossain

Excel Crash Course
for Engineers

 Springer

Eklas Hossain
Oregon Institute of Technology
Klamath Falls, OR, USA

ISBN 978-3-030-71038-5   ISBN 978-3-030-71036-1   (eBook)
https://doi.org/10.1007/978-3-030-71036-1

This Springer imprint is published by the registered company Springer Nature Switzerland AG
The registered company address is: Gewerbestrasse 11, 6330 Cham, Switzerland

Preface

Key Features of This Book

- The examples and exercises are related to engineering problems.
- Step-by-step demonstration of techniques.
- Every step includes vivid pictures for clear understanding.
- Contents cover important fields of engineering, espccially electrical engineering.
- Exercises at the end of the related topic, for easy referencing. Chap. 1 is the only exception with end-of-chapter exercise.

Microsoft® Excel has become an integral tool for many people. It is used by people in all fields, from students, teachers, researchers to businessmen, finance analytics, and many more. *Excel Crash Course for Engineers* presents the uses of MS Excel in engineering applications, particularly for electrical engineers. In this book, MS Excel is used in a new manner in the field of engineering, especially in electrical engineering. It demonstrates how a problem can be formulated and solved, how VBA is used, and how data is processed, cleaned, analysed, and visualized. This book is for people in technical fields, students and professionals alike. Its aim is to show the usefulness of Microsoft® Excel in solving a wide range of numerical problems. The book contains pictures of the step-by-step implementation of different functions of MS Excel that will help readers understand the concepts.

The book is comprised five chapters. Chapter One, 'Introduction to Microsoft Excel,' portrays the basics of MS Excel, starting with its background, purpose, and importance. The basic structure of Excel and its user interface is explained in detail, alongside discussing the basic keywords and terminologies used in Excel. Chapter Two, 'Graphing and Charting,' addresses the numerous types of graphs and charts that can be created in MS Excel, and demonstrates practical examples for each type. Chapter Three, 'Microsoft Excel Functions and Formulae,' clarifies the functions and formulae that Excel provides, and how they can be used to make tasks easier. Chapter Four, 'Microsoft Excel VBA,' describes the creation and use of user-defined functions in Excel. The concepts of Visual Basic for Applications (VBA) and macros are explicitly explained in this chapter. The final chapter is 'Microsoft Excel in Engineering Data and Analysis,' which highlights some very essential features in

MS Excel and includes many handy techniques to make Excel more user-friendly. The five chapters neatly summarize the entire concept, usage, and applications of Microsoft Excel.

The book has been written with engineers in mind and special focus has been made towards electrical engineers, with several electrical engineering problems being incorporated into the book. The book has been kept very simple and concise, careful not to confuse readers with technical jargon that might confuse readers. The author sincerely believes that *Excel for Engineers* will turn out to be a good read and a good investment of time and effort.

Klamath Falls, OR, USA   Eklas Hossain

Acknowledgements

The author expresses deep gratitude to everyone who contributed to making this book possible. All the information in this book has been borrowed from numerous freely available sources, and I merely aspired to accumulate the necessary information to benefit engineers who work with Microsoft Excel. In this regard, I would like to thank the millions of people who have developed and recorded these amazing methods that are valuable enough to document in a book such as this one.

I am sincerely grateful to my loving family, friends, well-wishers, colleagues, and my dear students for believing in me and helping me complete this book. Most importantly, I am indebted to my Creator for blessing me with a beautiful and healthy life which I strive to embellish with good deeds. This book is my humble attempt to serve humanity.

Contents

About the Author

Eklas Hossain (M'09, SM'17) received his PhD from the College of Engineering and Applied Science at the University of Wisconsin Milwaukee (UWM). He received his MS in mechatronics and robotics engineering from the International Islamic University of Malaysia, Malaysia, in 2010 and BS in electrical and electronic engineering from Khulna University of Engineering and Technology, Bangladesh, in 2006. Dr. Hossain has been working in the area of distributed power systems and renewable energy integration for the last 10 years, and he has published a number of research papers and posters in this field. He is now involved with several research projects on renewable energy and grid tied microgrid systems at Oregon Tech as an associate professor in the Department of Electrical Engineering and Renewable Energy since 2015. He is the senior member of the Association of Energy Engineers (AEE). Dr. Hossain is currently serving as an associate editor of *IEEE Access*. He is working as an associate researcher at Oregon Renewable Energy Center (OREC). Dr. Hossain is a registered professional engineer (PE) in the state of Oregon, USA. He is also a certified energy manager (CEM) and renewable energy professional (REP). His research interests include modelling, analysis, design, and control of power electronic devices; energy storage systems; renewable energy sources; integration of distributed generation systems; microgrid and smart grid applications; robotics; and advanced control system. He is the winner of the Rising Faculty Scholar Award in 2019 from the Oregon Institute of Technology for his outstanding contribution in teaching. Dr. Hossain, with his dedicated research team, is looking forward to exploring methods to make electric power systems more sustainable, cost-effective, and secure through extensive research, and analysis on energy storage, microgrid system and renewable energy sources.

Introduction to Microsoft Excel

1

Learning Objectives
- Learn about the history, purpose, and importance of Microsoft Excel
- Acquire knowledge of Microsoft Excel and its Graphical User Interface
- Discover spreadsheet and different applications of spreadsheet
- Acquaint with different components of Microsoft Excel
- Know about absolute and reference cells of Excel

1.1 Background of Microsoft Excel

1.1.1 History of MS Excel

Microsoft Excel is Microsoft's spreadsheet program created in 1993. The software is supported in macOS, Windows, iOS, and Android. It is used for calculations, graphical visualization of data, pivoted tables, and much more. It further uses Visual Basic for Application (VBA), which is a macro programming language. The history of the current day Excel is long, with gradual developments taking place in subsequent versions. The following table summarizes the historical development of MS Excel.

E. Hossain, *Excel Crash Course for Engineers*,
https://doi.org/10.1007/978-3-030-71036-1_1

Year	Version	Features
1982	Multiplan	Popular in CP/M systems, but got overtaken by Lotus 1-2-3 in MS-DOS systems
1985	Excel (for Mackintosh only)	Microsoft inaugurated the first version of Excel for Apple Inc.'s Mackintosh
1987	Excel 2.0 (for Windows)	Excel 2.0 had a heavy-graphics interface for Windows computers. Lotus was still incompatible with Windows, thus allowing Excel to top the market as a spreadsheet program.
1990	Excel 3.0	Excel 3.0 constituted toolbar, drawing, outlines, shortcuts, add-ins, 3D plots, etc.
1992	Excel 4.0	Excel 4.0 introduced autofill.
1993	Excel 5.0	Excel 5.0 started Visual Basic for Applications (VBA) or Macros.
1995	Excel 95 (v7.0)	As part of Office 95, Excel 95 was faster and built for the latest 32-bit computers using Intel's 386 microprocessor.
1997	Excel 97 (v8.0)	As part of Office 97, Excel 97 was a significant upgrade and first had the famous paperclip office assistant. The Natural Language labels were also introduced in this version, although removed later.
2000	Excel 2000 (v9.0)	As part of Office 2000, Excel 2000 was a minor upgrade. The clipboard was updated to hold multiple objects simultaneously. The paperclip assistant was made less intrusive.
2002	Excel 2002 (v10.0)	As part of Office XP, Excel 2002 contained minimal upgrades.
2003	Excel 2003 (v11.0)	As part of Office 2003, Excel 2003 also had very little news to offer but introduced the Tables feature.
2007	Excel 2007 (v12.0)	As part of Office 2007, Excel 2007 was a major upgrade and introduced the Ribbon menu, SmartArt, Name Manager, increased flexibility in graph formatting, upgrades in pivot tables, etc. The Office Open XML file formats were also introduced, denoting workbooks without macros as .xlsx and those with macros as .xlsm.
2010	Excel 2010 (v14.0)	Version 13 was skipped, and Excel 2010 was released as part of Office 2010. This version supported 64-bit computers and had improvements such as multi-threading recalculation (MTR), sparklines, image editing, backstage, customized Ribbon, etc.
2013	Excel 2013 (v15.0)	As part of Office 2013, Excel 2013 contained many new features, such as improved memory contention and multi-threading, power pivot, power view, flash fill, etc.
2016	Excel 2016 (v16.0)	As part of Office 2016, Excel 2016 contained many new features, such as read-only mode, new chart types, power query integration, forecasting functions, etc.
2019	Excel 2019 (v16.0)	Microsoft stopped releasing new versions of Excel and developed a system to incorporate upgrades to the features simply by Windows Update. The version remains 16.0.

1.1.2   Purpose of MS Excel

Microsoft Excel introduces us to a powerful analytic and data logging tool. It is a great program to represent data into tables and graphs. As data are presented in columns and rows, different mathematical operations can be performed for the purpose of converting data into information. Excel is very handy to deal with a huge amount of data. Excel allows easy handling, proper organization, and fast calculation with a large amount of data, which would otherwise be a very tedious job for a human. Excel is not confined within numerical data, it is also a great way to organize textual data, records, and lists. Almost all organizations make use of Excel to keep track of valuable information. Excel has made automation much easier. Official works which involved loads of files and paperwork can now be simply organized using Excel.

1.1.3   Importance of MS Excel

Research works deal with a lot of information, and requires organization and manipulation of a huge amount of data. Instead of doing this tedious task manually, various programming languages, spreadsheets, or other software can be used. MS Excel is the most common and user-friendly tool to handle large numbers of data. Both textual and numerical information can be contained in MS Excel, which makes it useful for numerous applications. The activities that can be accomplished in MS Excel can be done in many other applications as well. What makes Excel unique is it is an extremely simple and user-friendly interface that anyone would be comfortable to work with. Anyone with a basic computer knowledge can easily get acquainted with the various features in MS Excel by spending a minimum amount of time and effort.

1.2   Microsoft Excel Structure

1.2.1   Excel Graphical User Interface

Excel is a spreadsheet program based on menu and icon. The Graphical User Interface (GUI) is illustrated below. Some brief narratives for each menu item are also provided here. Microsoft Excel 2013 has been used in this book. The interface is described in brief as follows:

1. *Quick Access Toolbar*: It consists of options to save, undo, clear, or customize the toolbar.
2. *Ribbon Tab*: It is a strip of buttons consisting of menus, and almost all the functions available in Excel. It can be collapsed or expanded by double clicking in one of the options such as "HOME," or by pressing "CTRL+F1" in the keyboard.
3. *Name Tab*: It is a field that indicates the selected cell number. While selecting the data, it denotes the dimension of the data, i.e., the row and column of the table.
4. *Formula Bar*: An important field to use the formulae in Excel, denoted by "fx." By pressing "=" before the formula activates the field.
5. *Row and Column Tab*: The tabs enumerate the row and columns according to which the cells are named. The rows are named in numbers (1,2,3… about 1 mil-

lion) and the columns are named in letters (A, B, C...... AA, AB, AC, ... about 16,000).

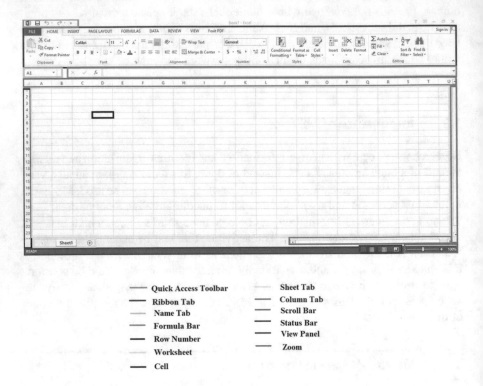

Quick Access Toolbar   Sheet Tab
Ribbon Tab   Column Tab
Name Tab   Scroll Bar
Formula Bar   Status Bar
Row Number   View Panel
Worksheet   Zoom
Cell

Fig. 1.1

6. *Worksheet*: One page in a workbook is called a worksheet, where the data can be added or manipulated. They are divided into several unit boxes called cells with rows and columns numbered in row and column tab.

7. *Cell*: The small boxes inside the worksheet are known as cells. The cells can contain numbers, text, images, equations, and many more. There are about 17,179,869,184 cells available in the latest version of Microsoft Excel. They are named merging the name of the column and row, for example, A1, B4, D6, etc.

8. *Sheet Tab*: This tab is used to open a new worksheet under the same workbook. Each worksheet has the same features. The number of sheets that can be opened depends upon the memory of the computer.

9. *Scroll Bar*: Two scroll bars, one for horizontal view, another for vertical movement exists for Microsoft Excel.

10. *Sheet Tab*: This tab is used to open a new worksheet under the same workbook. Each worksheet has the same features. The number of sheets that can be opened depends upon the memory of the computer.

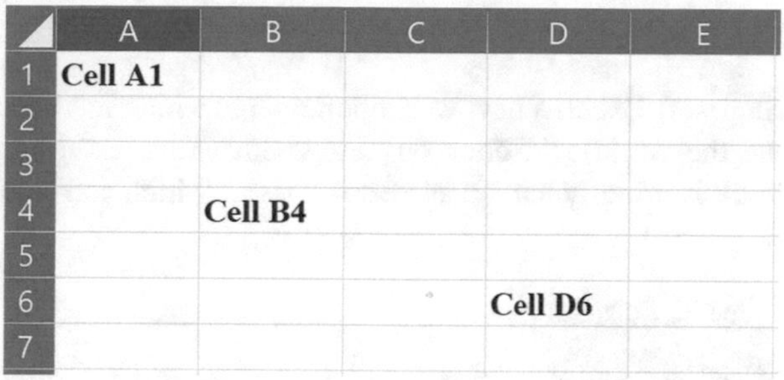

	A	B	C	D	E
1	**Cell A1**				
2					
3					
4		**Cell B4**			
5					
6				**Cell D6**	
7					

Fig. 1.2

11. *Scroll Bar*: Two scroll bars, one for horizontal view, another for vertical movement exists for Microsoft Excel.
12. *Status Bar*: This has multiple functionalities as can be seen by right clicking on the bar. It usually notifies the status of the Excel file that is being used. For example, if it is ready to use, or busy to calculate data. It also shows numerical count, data average, or summation based on the data selected and several other functionalities.
13. *View Panel*: There are usually three options available to change the view of the worksheet. "Normal" provides a default view of the whole worksheet. "Page Layout" splits the screen into segments. "Page Break Preview" provides a zoomed-out view of the worksheet holding more cells.
14. *Zoom*: A scroll bar to zoom in or out from the worksheet. The default size is 100% in the middle of the bar.

1.2.2   Workbooks and Worksheets

When an Excel file is opened, it displays a "Book" with numerous cells in it. The page shown in Fig. 1.1 is a "Workbook" in Excel. But sometimes, it might be necessary to create several tables so that they can all be used for the same purpose. For example, if one wants to create several tables of data and want to use some of the data in a master table, one will need to search individually and place those data in the master table for calculation, which is undoubtedly a tedious task. The rafts of data may cause format loss and can cost a large amount of time. Microsoft Excel includes such a tool, called "Worksheet," which can create multiple tables in a single workbook, so that their data can be called and used from any sheet. This can be analogical to the pages of the books, where the book per se is the "Workbook" and the pages are the "Worksheet."

1.2.3   Opening and Saving Workbooks and Worksheets

On opening Microsoft Excel, a new workbook opens, which has one worksheet by default. To save the workbook, click on "Files>Save/Save as" to any location of choice. Workbook is initially named as "Book1.xlsx," which can be renamed while saving the file.

Fig. 1.3

   To add a new sheet, press the small "Plus" symbol at the Sheet Tab. Figure 1.4 shows five sheets. The new sheet will replace its previous sheet, and the previous sheet can reappear upon clicking the arrow or dots beside the tab.

Fig. 1.4

   Each of these tabs can be manipulated by right-clicking onto the tab. The tabs can be renamed, colored, hidden, and even protected, by locking any particular attribute of the worksheet. As all of these worksheets are associated with a single workbook, saving the workbook will save all of the sheets.

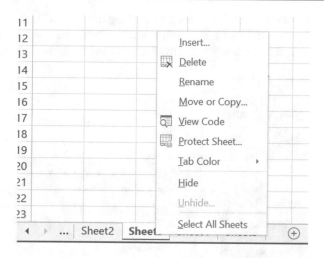

Fig. 1.5

To start a new workbook, click "Files>New." A variety of templates may be available depending on the version, which includes designs and models for calendar, budget, invoice, report, and others. To work on a raw Excel file, click "Blank workbook" which would open a new workbook with a single worksheet in a new window.

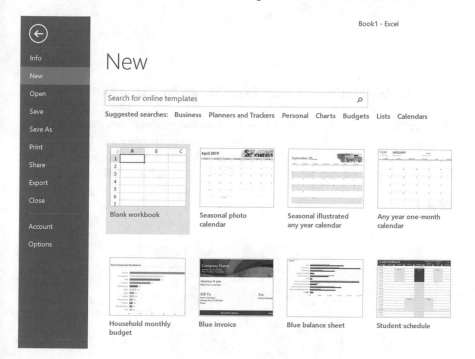

Fig. 1.6

An existing workbook can be opened by clicking on "Files>Open" which would lead to the computer folders for navigation. On browsing the appropriate location, or by searching in the "Recent workbooks," the desired Excel file can be opened.

Fig. 1.7

1.2.4   Consolidating Multiple Excel Sheets Into One Sheet

Consolidating two sheets means keeping the title and headers from two Excel sheets same, and conducting mathematical operations upon two entries in the existing or another sheet. This is useful for large tables where it is tedious to use formulae. Consider two excel sheets with the prices of five food items as follows:

Sheet1:

	A	B	C
1	Item	Price	
2	Syrup	$10	
3	Sugar	$17	
4	Vanilla	$23	
5	Cocoa	$57	
6	Milk	$20	
7			

Sheet1   Sheet2   Sheet3

Fig. 1.8

Sheet2:

	A	B
1	Item	Price
2	Syrup	$45
3	Sugar	$36
4	Vanilla	$15
5	Cocoa	$85
6	Milk	$30
7		

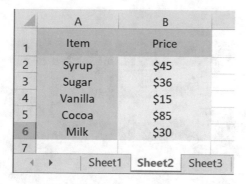

Fig. 1.9

In order to consolidate these two sheets into a new sheet named "Sheet3," do the following:

1. Check that the headers or titles for each sheet are the same, and there are no missing data in the tables.
2. Create a new sheet "Sheet3" and go to "Data">"Consolidate."

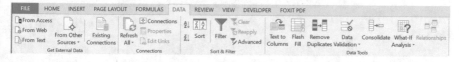

Fig. 1.10

A new pop-up window will appear.

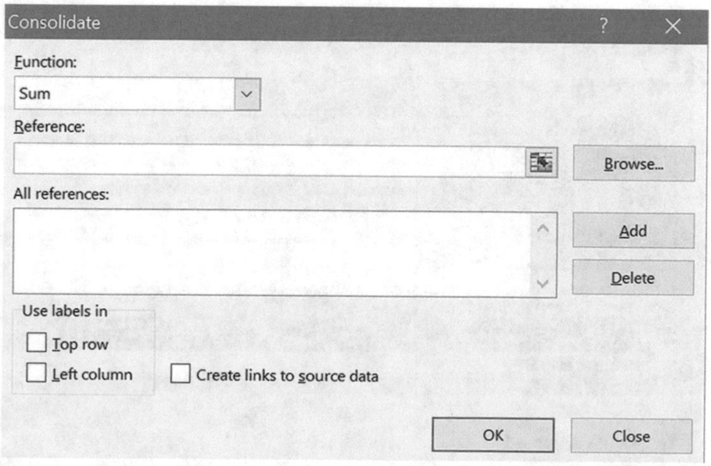

Fig. 1.11

3. Consider that we want to add two values of the price for each item. Select the function "Sum" to perform the operation. There are other options available, as shown below:

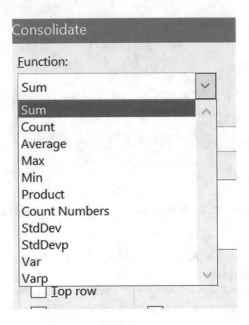

Fig. 1.12

4. In the "Reference" box, the sheets are needed to be added. The sheets can be browsed from a different location, or existing tables can be selected for referencing using the button at the end. Pressing the button will collapse the window for letting the user select the table array.

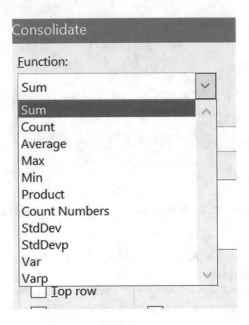

Fig. 1.13

5. Again, click on the button, the window will return to its full form. By pressing "Add," the reference will be added to the box "All references" as shown below.

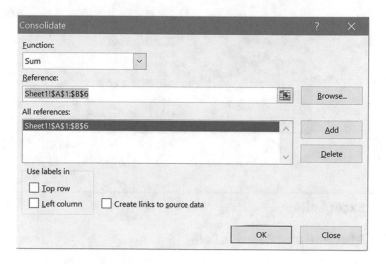

Fig. 1.14

6. Similarly, add the table in Sheet2.

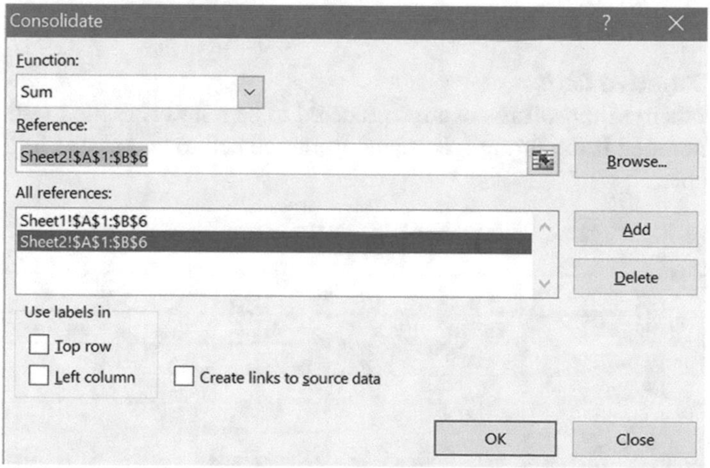

Fig. 1.15

Click OK.

7. In sheet3, it is seen the added data are automatically added.

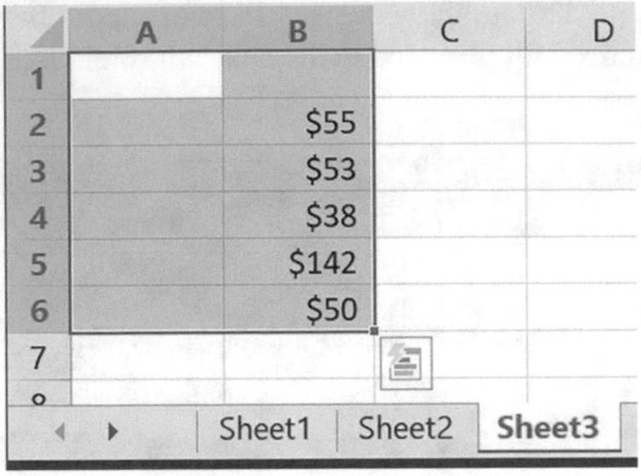

	A	B	C	D
1				
2		$55		
3		$53		
4		$38		
5		$142		
6		$50		
7				

Sheet1   Sheet2   **Sheet3**

Fig. 1.16

1.3   Excel Cells

1.3.1   Absolute and Relative Cell

In MS Excel, the cells can be categorized into two types, absolute and relative. Both of them have distinct properties, and this is why they behave differently when copied to another cells. Absolute references are "absolute," that means they do not change wherever they are copied. On the other hand, relative cells are subject to change if a formula is copied to another cell.

1.3.1.1 Relative Cells

All the cells in Microsoft Excel are considered to be relative, as the location of the cells is modified if the formula is copied from one cell to and pasted into another. For example:

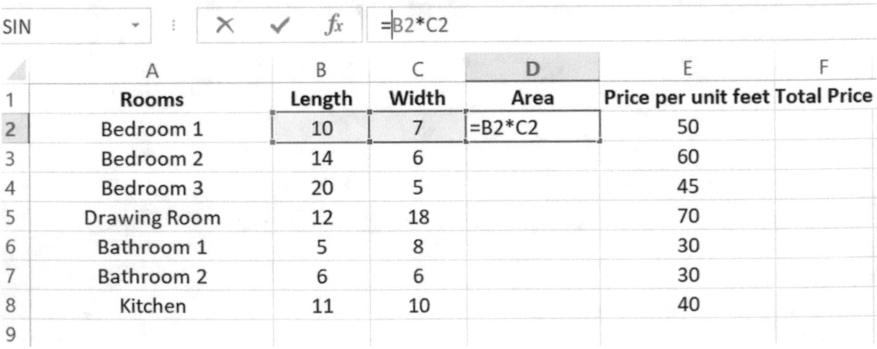

SIN	f_x =B2*C2

	A	B	C	D	E	F
1	**Rooms**	**Length**	**Width**	**Area**	**Price per unit feet**	**Total Price**
2	Bedroom 1	10	7	=B2*C2	50	
3	Bedroom 2	14	6		60	
4	Bedroom 3	20	5		45	
5	Drawing Room	12	18		70	
6	Bathroom 1	5	8		30	
7	Bathroom 2	6	6		30	
8	Kitchen	11	10		40	
9						

Fig. 1.17

In the example shown, the area of each of the room is to be determined. Now as each of the cell works as a relative cell by default, let us put the formula for the area of Bedroom 1. For writing the formula for multiplication, the following format is used:

$$= (\text{Cell number of the first value}) * (\text{Cell number of the second value}).$$

In this case, this is:

$$= B2 * C2$$

The formula can either be written in the resultant cell (in this case, D2) or in the formula bar by selecting that particular cell.

Press "Enter" to get the result. The result will appear. Now, for other cells in the D column, instead of typing the formula for each cells, the same formula (=B2*C2) can be extended for the other cells. It is because as the cells are relative, on copying the formula in other cells, the number of the cells will modify automatically to address the corresponding cells. For example, for cell D3, the formula will be:

$$= B3 * C3$$

For each cell, retyping the formula is cumbersome. The "file handler" of the selected cell at the bottom right corner (marked in red in the figure) takes the advantage of the relative cell, and replaces the corresponding cell number maintaining the same formula.

D2	▾ ⋮	× ✓	f_x	=B2*C2		
◢	A	B	C	D	E	F
1	Rooms	Length	Width	Area	Price per unit feet	Total Price
2	Bedroom 1	10	7	70	50	
3	Bedroom 2	14	6		60	
4	Bedroom 3	20	5		45	
5	Drawing Room	12	18		70	
6	Bathroom 1	5	8		30	
7	Bathroom 2	6	6		30	
8	Kitchen	11	10		40	
9						

Fig. 1.18

To apply a similar function for the whole column, click the "file handler" and drag it until D8 to get the same formula with replaced cell number.

	A	B	C	D	E	F
1	Rooms	Length	Width	Area	Price per unit feet	Total Price
2	Bedroom 1	10	7	70	50	
3	Bedroom 2	14	6		60	
4	Bedroom 3	20	5		45	
5	Drawing Room	12	18		70	
6	Bathroom 1	5	8		30	
7	Bathroom 2	6	6		30	
8	Kitchen	11	10		40	

Fig. 1.19

	A	B	C	D	E	F
1	Rooms	Length	Width	Area	Price per unit feet	Total Price
2	Bedroom 1	10	7	70	50	
3	Bedroom 2	14	6	84	60	
4	Bedroom 3	20	5	100	45	
5	Drawing Room	12	18	216	70	
6	Bathroom 1	5	8	40	30	
7	Bathroom 2	6	6	36	30	
8	Kitchen	11	10	110	40	
9						

Fig. 1.20

On double clicking on any of the cells in the column, the formula will be visible.

SIN	▾	:	✕ ✓	*fx*	=B5*C5	

	A	B	C	D	E	F
1	Rooms	Length	Width	Area	Price per unit feet	Total Price
2	Bedroom 1	10	7	70	50	
3	Bedroom 2	14	6	84	60	
4	Bedroom 3	20	5	100	45	
5	Drawing Room	12	18	=B5*C5	70	
6	Bathroom 1	5	8	40	30	
7	Bathroom 2	6	6	36	30	
8	Kitchen	11	10	110	40	
9						

Fig. 1.21

1.3.1.2 Absolute Cells

In contrast to relative cells, absolute cells do not change the cell numbers in their formulae. For example, in the illustration shown for relative cells, if one wants to design all of the rooms with the price of the first bedroom only, instead of individual values, the area will just be multiplied by the value of cell E2. Then all other values

(E3 to E8) will not be necessary for the purpose. To denote a cell as an absolute cell or a reference cell, a dollar ($) sign is used preceding the number. For example, to set both the column and row of cell E2 to an absolute one, type $E$2. Similarly, to make the row unchanged, type E$2. $E2 indicates that the column would not be changed while typing the formula. Here, in the above formula, to fill the total cost with the price of E2, the formula should be typed in F2 or in the formula bar,

$$= D2 * \$E\$2$$

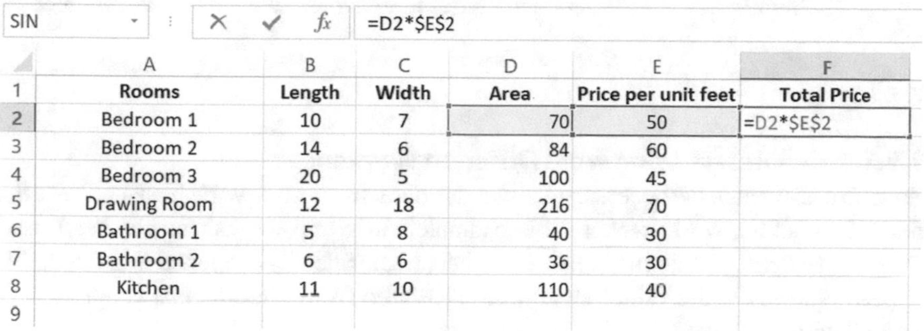

SIN		⋮	✕	✓	fx	=D2*$E$2	

	A	B	C	D	E	F
1	**Rooms**	**Length**	**Width**	**Area**	**Price per unit feet**	**Total Price**
2	Bedroom 1	10	7	70	50	=D2*$E$2
3	Bedroom 2	14	6	84	60	
4	Bedroom 3	20	5	100	45	
5	Drawing Room	12	18	216	70	
6	Bathroom 1	5	8	40	30	
7	Bathroom 2	6	6	36	30	
8	Kitchen	11	10	110	40	
9						

Fig. 1.22

By pressing "Enter," the value becomes visible. Similarly, by using the file handler, the formula can be copied from cell E3 to E8, where the value of the rows in D column will change, keeping the price per unit feet the same ($E$2 being 50 every time).

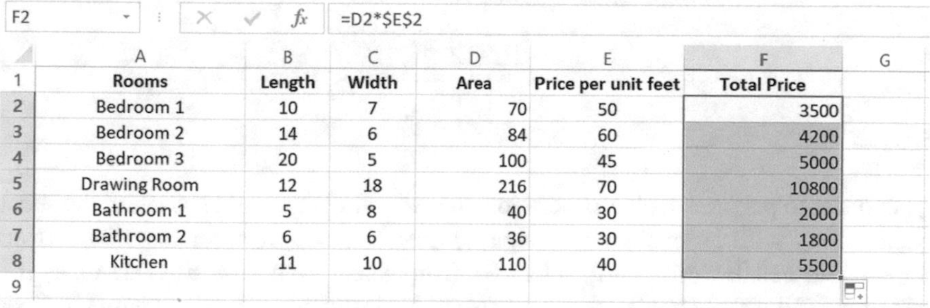

F2		⋮	✕	✓	fx	=D2*$E$2	

	A	B	C	D	E	F	G
1	**Rooms**	**Length**	**Width**	**Area**	**Price per unit feet**	**Total Price**	
2	Bedroom 1	10	7	70	50	3500	
3	Bedroom 2	14	6	84	60	4200	
4	Bedroom 3	20	5	100	45	5000	
5	Drawing Room	12	18	216	70	10800	
6	Bathroom 1	5	8	40	30	2000	
7	Bathroom 2	6	6	36	30	1800	
8	Kitchen	11	10	110	40	5500	
9							

Fig. 1.23

The formula in other cells can be checked by double clicking on any particular cell. For example, by clicking on cell E5:

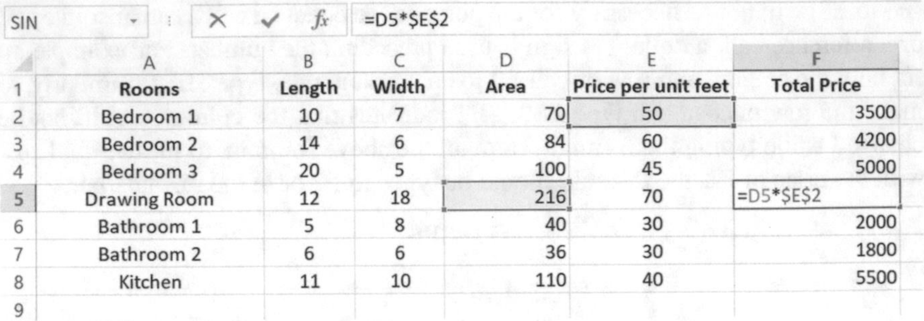

| SIN | ▾ | : | ✕ | ✓ | *fx* | =D5*$E$2 | |

	A	B	C	D	E	F
1	**Rooms**	**Length**	**Width**	**Area**	**Price per unit feet**	**Total Price**
2	Bedroom 1	10	7	70	50	3500
3	Bedroom 2	14	6	84	60	4200
4	Bedroom 3	20	5	100	45	5000
5	Drawing Room	12	18	216	70	=D5*$E$2
6	Bathroom 1	5	8	40	30	2000
7	Bathroom 2	6	6	36	30	1800
8	Kitchen	11	10	110	40	5500
9						

Fig. 1.24

1.3.1.3 Cell Referencing from Different Worksheet

Sometimes it may be necessary to use the data from one worksheet and use the result in another worksheet. In this example, only one worksheet has been used (Sheet1) to demonstrate the cells. If another sheet (Sheet2) is opened and two fields appear as follows, the values should be calculated from Sheet1 to be printed in B1 and B2 cells of Sheet2.

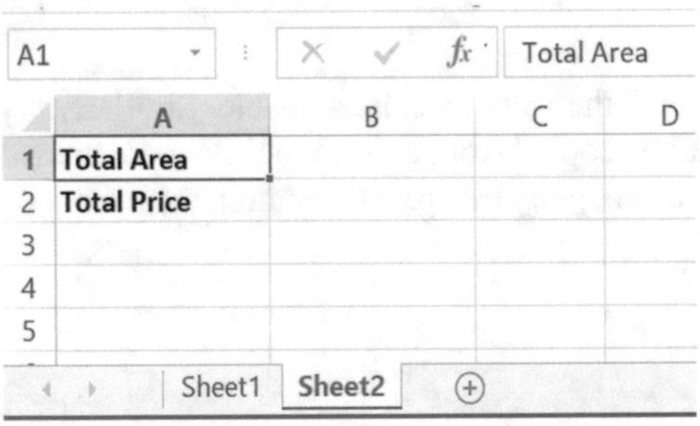

| A1 | ▾ | : | ✕ | ✓ | *fx* | Total Area | |

	A	B	C	D
1	Total Area			
2	Total Price			
3				
4				
5				

◀ ▶ | Sheet1 | **Sheet2** | ⊕

Fig. 1.25

In order to perform such an operation, the cell number should start with the sheet name, along with an exclamatory sign (!). For instance, to refer to cell A1 from Sheet1 into cell B1 of Sheet2, select B1 from Sheet 2 and use **Sheet1!A1** as the reference. If the sheet name contains a space, for instance, if the sheet name is **Task Sheet**, the cell can be referred with a single quotation mark ('), the reference being **'Task Sheet'!A1.**

As an example, in Sheet1, if the total area and cost are determined in cell D9 and F9 respectively using the following formula:

$$= SUM(\text{cell number the value starting from : cell number the value is ending at}).$$

For area, the formulation will be:

| SIN | ▾ | ⋮ | ✕ | ✓ | *fx* | =SUM(D2:D8) |

	A	B	C	D	E	F
1	**Rooms**	**Length**	**Width**	**Area**	**Price per unit feet**	**Total Price**
2	Bedroom 1	10	7	70	50	3500
3	Bedroom 2	14	6	84	60	4200
4	Bedroom 3	20	5	100	45	5000
5	Drawing Room	12	18	216	70	10800
6	Bathroom 1	5	8	40	30	2000
7	Bathroom 2	6	6	36	30	1800
8	Kitchen	11	10	110	40	5500
9				=SUM(D2:D8)		

Fig. 1.26

And for cost, the formulation will be:

| F9 | ▾ | ⋮ | ✕ | ✓ | *fx* | =SUM(F2:F8) |

	A	B	C	D	E	F
1	**Rooms**	**Length**	**Width**	**Area**	**Price per unit feet**	**Total Price**
2	Bedroom 1	10	7	70	50	3500
3	Bedroom 2	14	6	84	60	4200
4	Bedroom 3	20	5	100	45	5000
5	Drawing Room	12	18	216	70	10800
6	Bathroom 1	5	8	40	30	2000
7	Bathroom 2	6	6	36	30	1800
8	Kitchen	11	10	110	40	5500
9				656		32800
10						

Fig. 1.27

Chapter 3 will provide more insights on Excel functions and formulae. However, as the total area is 656 unit (cell D9) and the total cost is 32,800 unit (cell F9), these values have to be referred to cell B1 and B2 of Sheet2 respectively.

Refer the cells of B1 and B2 as **=Sheet1!D9** and **=Sheet1!F9** respectively.

| B2 | ▾ | ⋮ | ✕ | ✓ | *fx* | =Sheet1!F9 |

	A	B	C	D
1	**Total Area**	656		
2	**Total Price**	32800		
3				
4				
5				

◄ ► | Sheet1   **Sheet2**   ⊕

Fig. 1.28

Did you check the formula shown in the formula bar for cell B2? The formulations will automatically modify themselves if the sheet names are changed after setting up the formula.

1.4   Conclusion

This chapter introduces Microsoft Excel and its basic concepts. Readers will be able to get acquainted to the basic interface offered by MS Excel and know the features readily available in Excel. The concepts of workbook, worksheet, cells, rows, and columns will be clear after reading this chapter. The distinction between absolute and reference cells will also be understood. The method of cell referencing is also explained here. This chapter will be helpful for novice users of MS Excel and provide the preliminary training on this spreadsheet program.

Exercise 1

1. What are workbook and worksheets?
2. What is the difference between absolute and reference cells?
3. State five uses of Microsoft Excel.
4. Open a new workbook in Excel and write down the times table of 247 using the techniques learnt in this chapter. Save the file as "Times Table of 247" in your Desktop.
5. A school is raising funds for helping the treatment of a critically ill student named Sarah. In an Excel worksheet, write the names of 30 students and record how much money they paid for the cause. Find the total money accumulated. In the same workbook, start a different worksheet and prepare a similar list of the fund gathered from 20 staff members and teachers. In a third worksheet of the same workbook, find out the total amount collected from all sources mentioned. Save the workbook as "Fund Collected for Sarah."

Graphing and Charting

2

Learning Objectives
- Discover different charts available in Microsoft Excel
- Visualize data with different charts
- Manipulate designs and properties of charts
- Export images from Microsoft Excel
- Know the practical uses of Excel charts and graphs

2.1 Line Chart

Line charts present the change of data with time. We know them as graphs. Plotting graphs is a useful feature in Excel. A line chart is used to insert numeric labels, text labels, or time stamps on the horizontal axis. Line charts may be two or three-dimensional.

2D Plot:

1. *Line or Line with Marker* – used to show the trend over time.
2. *Stacked line or Stacked line with Marker* – used to show the share of the elements to a whole.
3. *100% Stacked line or 100% Stacked line with Marker* – used to show the share of the elements based on their percentage; suitable for a large amount of data.

© The Author(s), under exclusive license to Springer Nature Switzerland AG 2021
E. Hossain, *Excel Crash Course for Engineers*,
https://doi.org/10.1007/978-3-030-71036-1_2

3D Plot:
This chart is used to show data based on three axes and their changes with respect to categories.

Marked Line Charts: The marked line chart marks the data points; thus, it is suitable for charts where a limited number of data is available. For a large number of data, the marked line charts should be avoided. To generate a marked line chart, carry out the following steps.

Suppose a dataset for the engineering students getting admitted to a university is provided as follows:

Year	Electrical and Electronic Engineering	Computer Science and Engineering	Biomedical Engineering
2015	303	256	143
2016	343	289	165
2017	377	353	178
2018	397	321	172
2019	419	311	186

1. Type or copy the data in the Excel file. To copy, select the above data and copy this to save in the buffer memory. Then go to Excel, select the cell from which the table should begin, and paste the data. The spreadsheet should look as follows.

	A	B	C	D
		Electrical and Electronic Engineering	Computer Science and Engineering	Biomedical Engineering
1				
2	2015	303	256	143
3	2016	343	289	165
4	2017	377	353	178
5	2018	397	321	172
6	2019	419	311	186
7				

Fig. 2.1

2. Go to Insert > Charts. Select the "Insert Line Chart" symbol.

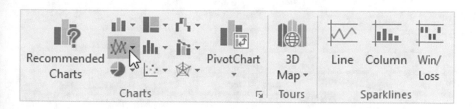

Fig. 2.2

3. There are several options available. To show the change in trend with markers, click "Line with Markers."

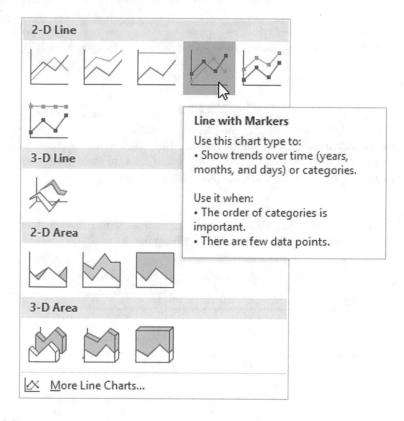

Fig. 2.3

Result:

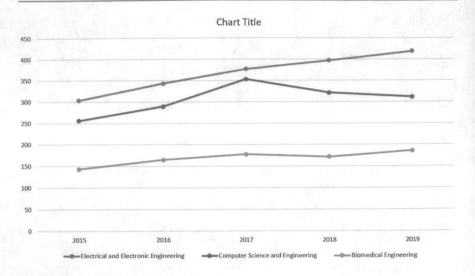

Fig. 2.4

In the case of numeric labels, ensure that cell A1 is empty before creating the line chart. Thus, Excel cannot identify the entries in column A as a data series and puts the data on the horizontal axis automatically. After the line chart is created, the text "Year" can be inserted into cell A1. Follow Section 2.1.1. to learn changing the axis type, adding titles to the axes, and changing the axis scale.

Exercise 2.1:   Make a data table and plot this line chart with a marker from below. Also, draw the stacked line and 3D line chart from the data.

Month	Product-A	Product-B	Product-C	Product-D
Jan	23	47	14	64
Feb	12	53	5	48
Mar	16	33	21	47
Apr	27	34	15	45
May	9	40	18	39
Jun	20	46	10	53

2.1.1   Line Chart Properties

As soon as the chart appears, the chart elements can be accessed and modified from the + sign at the upper right corner of the chart window. The style and color of the chart can be changed from the second option. By clicking on different segments of the chart, the chart title, axis title, legend texts, and legend positions can be modified.

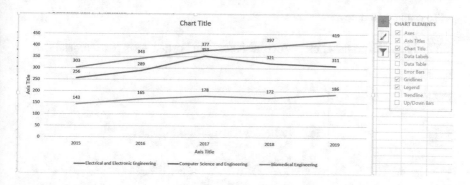

Fig. 2.5

2.1.1.1 Axis Type

The format of the axes can be changed from text axis to date axis in case the date needed to be presented in the horizontal axis is not in sequential order. Even though Excel automatically selects the text or date axis, sometimes it may be required to change the type manually. This rule is not applicable for dates in legend. For the following dataset:

Date	Number of visitors
11/04/2019	255
12/04/2018	363
13/07/2015	312
14/11/2019	415
15/01/2019	278

After plotting the line chart, this should look like the following for the text axis:

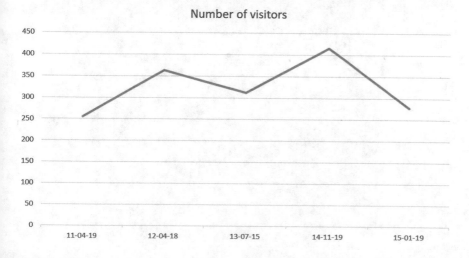

Fig. 2.6

To convert to date axis:

1. On the x-axis, right-click > **Format Axis**. The Format Axis pane slides in.

Fig. 2.7

2. Click Text axis.

Fig. 2.8

Result:

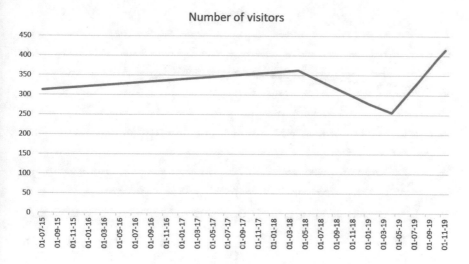

Fig. 2.9

This function is available for line charts, area charts, bar charts, and column charts.

2.1.1.2 Axis Titles
For adding a title on the vertical axis, select the chart and follow these steps:

1. Clicking on the + button beside the chart, select Axis Titles > Primary Vertical.

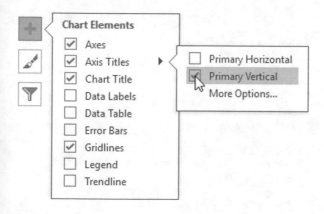

Fig. 2.10

2. Give a suitable title to the vertical axis. For the above example, the title is changed to "Visitors."

Result:

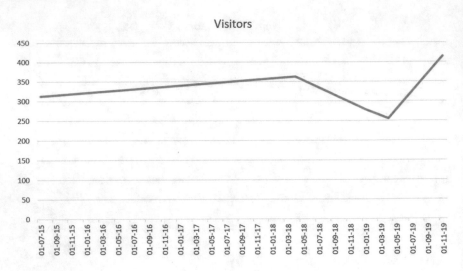

Fig. 2.11

2.1.1.3 Axis Scales

The values of the vertical axis are automatically determined by Excel. To manually alter these values, follow this process:

1. Right-click on the vertical axis. Click on Format Axis. Then the Format Axis pane pops up.

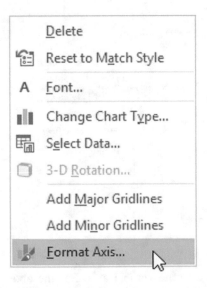

Fig. 2.12

Fig. 2.13

2. Set the Maximum = 1000 for the bound and Major = 100 for the units.

Result:

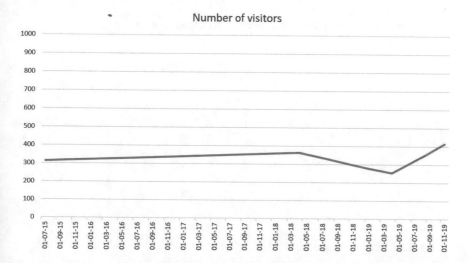

Fig. 2.14

2.1.2 Error Bars

The following example shows the method of adding error bars to a line chart in Excel. Consider the following data:

Month	Forecasted	Predicted
Sun	32	36
Mon	45	39
Tue	22	21
Wed	53	50
Thu	63	67

1. Copy the chart in Excel and select the data.

	A	B	C
1	**Month**	**Forecasted**	**Predicted**
2	Sun	32	36
3	Mon	45	39
4	Tue	22	21
5	Wed	53	50
6	Thu	63	67
7			

Fig. 2.15

2. Select Insert > Line chart. The line chart will look like this:

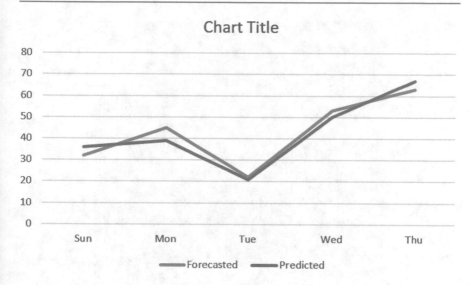

Fig. 2.16

3. From the + button beside the chart, go to Error Bars > More Options.

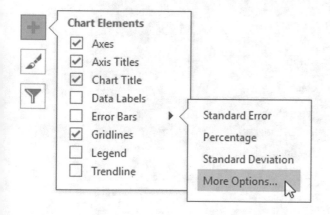

Fig. 2.17

There is a shortcut to display error bars using the Standard Error, which shows a percentage value of 5% or one standard deviation.

The Format Error Bars pane appears.

4. Choose a Direction. Click Both.
5. Choose an End Style. Click Cap.
6. Click on Fixed value. Enter the value 10.

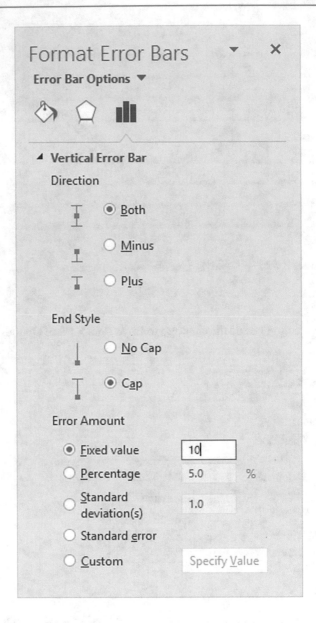

Fig. 2.18

Result:

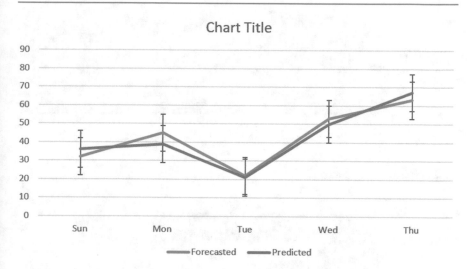

Fig. 2.19

2.2   Column Charts

Column charts can compare values of different categories by the use of adjacent vertical bars. Like line charts, column charts can also be represented in 2D or 3D Line, each having three categories: stacked column, clustered column, and 100% stacked column. From the previous example, we want to create a clustered column chart for the students of the Computer Science and Biomedical Engineering Department. We see that the Year data and the student's data are not adjacent. To create the chart in this case, implement the following steps:

1. Select the cells A1:A6. Press and hold CTRL. Select the cells C1:C6 and then D1:D6, and let go of CTRL.

	A	B	C	D	E
1		Electrical and Electronic Engineering	Computer Science and Engineering	Biomedical Engineering	
2	2015	303	256	143	
3	2016	343	289	165	
4	2017	377	353	178	
5	2018	397	321	172	
6	2019	419	311	186	
7					
8					

Fig. 2.20

2. Select Insert > Charts > Column.

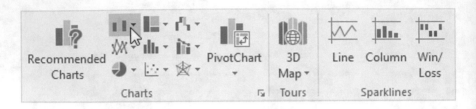

Fig. 2.21

3. Click Clustered Column.

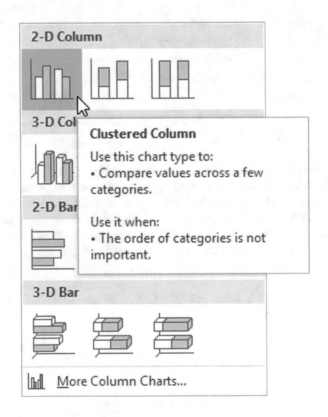

Fig. 2.22

Result:

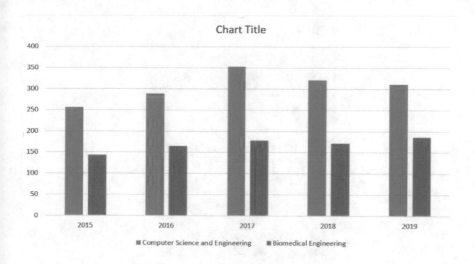

Fig. 2.23

The style of the graph and the color of the charts can be changed from the following ribbon (can be opened by "Chart Tool>Design" while the Chart is on the screen).

Fig. 2.24

A series of numbers plotted in a chart is termed a *data series*. A chart may have one or more data series plotted as above.

2.2.1   Data Source Box

Data Source boxes are used to manipulate the data, chart, format the axis, and for many other important tasks. To find the dialog box Select Data Source to switch axes, select the chart and follow this process:

1. Right-click > Select Data.

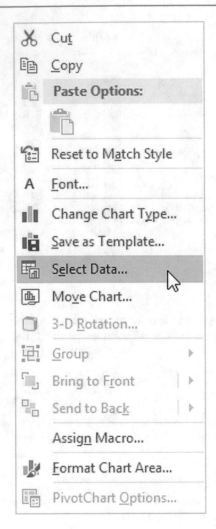

Fig. 2.25

The dialog box "Select Data Source" pops up.
2. On the left, the three data series of three departments are shown. On the right, the horizontal axis labels are shown.

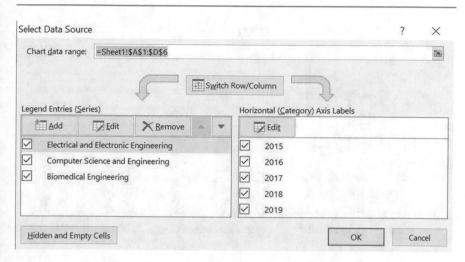

Fig. 2.26

3. If Switch Row/Column is clicked, there will be five data series (2015, 2016, 2017, 2018, and 2019) and three horizontal axis labels (three departments).

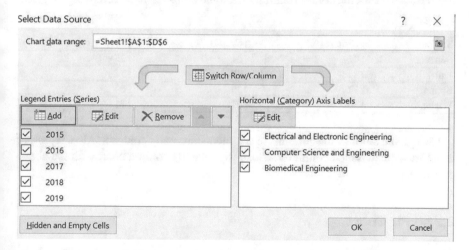

Fig. 2.27

Result:

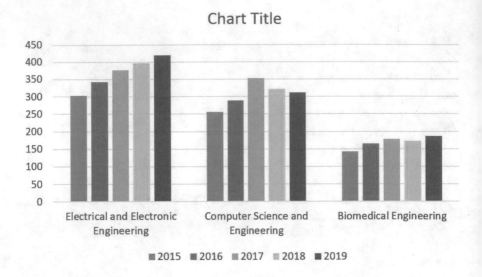

Fig. 2.28

Exercise 2.2: Plot the column chart for the three crops from the following diagram.

Seasons	Crop A (%)	Crop B (%)	Crop C (%)
Summer	45	51	25
Fall	34	60	37
Spring	17	39	64
Winter	30	22	71

Change the color of the columns to any monochromatic color set. Also, represent the chart with switched columns. Also, design the 3D column chart with the above properties.

2.3   Pie Chart

Pie charts visualize how much a particular factor contributes to a complete whole. For example, a pie chart can demonstrate the specific share of coal, oil, and natural gas in the total amount of fossil fuels used. Only one data series can be represented through a pie chart. Pie charts have the following categories:

2D Pie:

1. *Pie*: Regular pie chart to show the parts of a whole in a single pie chart.

2. *Pie of Pie*: If there are several slices that sum up to be less than 10%, it will be harder to distinguish. Pie of Pie expand those small portions into another secondary pie.

3. *Bar of Pie*: Used for a similar purpose, but in this chart, the small portions are shown as bar charts.

3D Pie: This chart is the 3D representation of the 2D Pie chart.

Doughnut: Doughnut charts are used instead of a pie chart in cases where multiple series are present that sums up to a larger whole.

To create a 2D pie chart following the previous example for the year 2015, execute this procedure:

1. Select the cells A1:D2.

	A	B	C	D	E
1		**Electrical and Electronic Engineering**	**Computer Science and Engineering**	**Biomedical Engineering**	
2	2015	303	256	143	
3	2016	343	289	165	
4	2017	377	353	178	
5	2018	397	321	172	
6	2019	419	311	186	
7					

Fig. 2.29

2. Go to Insert > Charts > Pie.

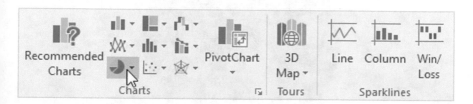

Fig. 2.30

3. Select Pie.

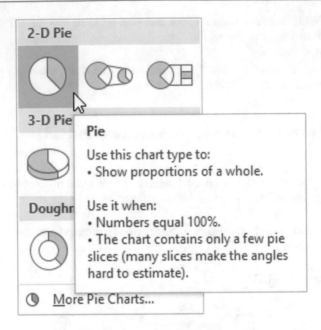

Fig. 2.31

Result:

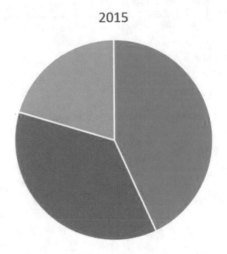

Fig. 2.32

4. At first, select the whole pie. Then click on any specific slice and move it from the center.

Result:

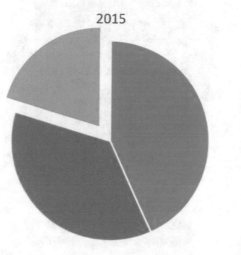

▪ Electrical and Electronic Engineering   ▪ Computer Science and Engineering   ▪ Biomedical Engineering

Fig. 2.33

Exercise 2.3: In a class of 60 students, 23 like football, 13 like basketball, 10 like baseball, 8 prefer swimming, and the rest like ice hockey. Plot the pie chart from this information.

2.3.1 Pie Chart Properties

Four properties of pie charts are narrated below:

(i) The data in the pie chart can be labeled in two ways, by adding data labels or by adding callout. By right-clicking above the pie, choosing "Add Data Labels" will add either of the labels. The image in Fig. 2.34a is a data label, and that in Fig. 2.34b is data callout.

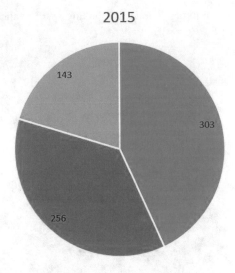

Fig. 2.34(a)

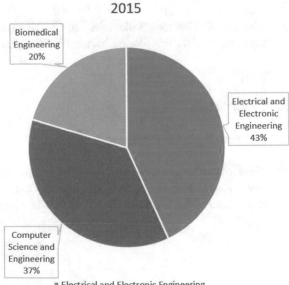

Fig. 2.34(b)

(ii) The representation of the data can be formatted from the following window by right-clicking on the chart > Format Data Labels.

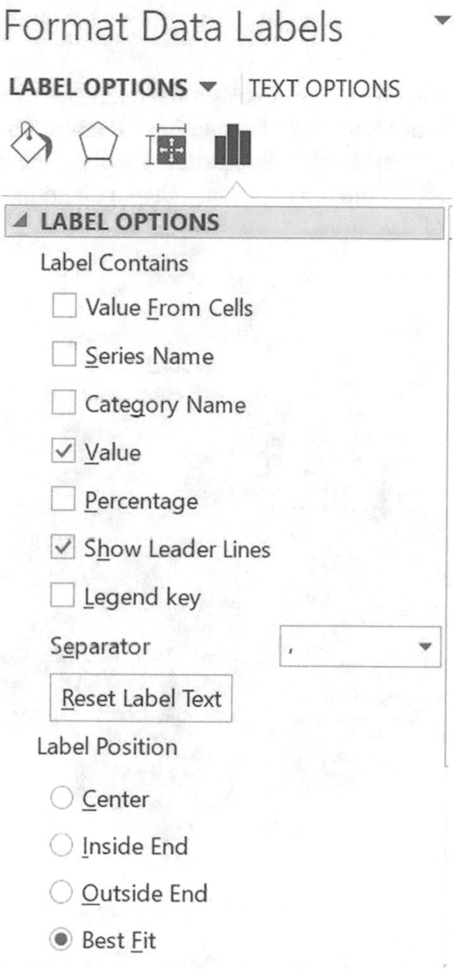

Format Data Labels ▾

LABEL OPTIONS ▾ TEXT OPTIONS

◇ ⬠ ▦ �░

◢ LABEL OPTIONS

Label Contains

☐ Value From Cells

☐ Series Name

☐ Category Name

☑ Value

☐ Percentage

☑ Show Leader Lines

☐ Legend key

Separator   [,   ▾]

[Reset Label Text]

Label Position

○ Center

○ Inside End

○ Outside End

◉ Best Fit

Fig. 2.35

(iii) To change the colors of the pie charts, while the pie chart is on the screen, select "Design" option from the "Chart Tool" and select "Change Colors." The background can also be changed from "Chart Styles." The ribbon will look as follows:

Fig. 2.36

(iv)   To rotate a pie chart, right-click on the chart > Format Data Series. Then, change the "Angle of First Slice" for rotating the chart. The size of the slice can also be controlled from the "Pie Explosion" section. The diagram in Fig. 2.37a shows a 100° rotation from the previously shown diagram, and Fig. 2.37b shows a 50% explosion of the slices.

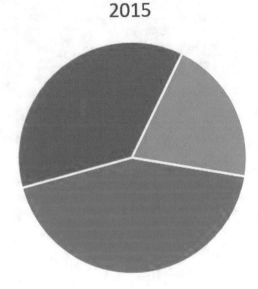

Fig. 2.37(a)

2015

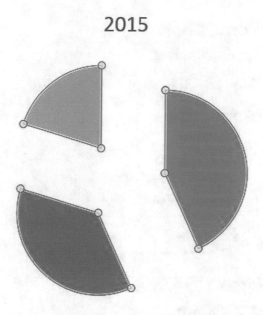

- Electrical and Electronic Engineering
- Computer Science and Engineering
- Biomedical Engineering

Fig. 2.37(b)

These values can be varied from the slider or by adding values in the text field.

Fig. 2.38

2.4   Gauge Chart

A *gauge chart* combines a Pie chart and a Doughnut chart. It is also termed a Speedometer chart due to its resemblance to a car's speedometer. Let us consider the following data:

Speedometer		Pointer	
Start	0	Value	70
Initial	15	Pointer	1

(Continued)

Speedometer		Pointer	
Middle	45	End	129
End	40		
Max	100		

The gauge chart will be a semicircle chart, where a black pointer will denote the value. Unlike other charts, the gauge chart is to be manually made, and the data needs further scaling and mapping to display in the chart. In fact, 50% of the Doughnut chart is used to present the data, and 1% of the pie chart is used to design the pointer. The methods to design the gauge chart are mentioned first.

Chart 1	▼	:	✕	✓	*fx*	

◢	A	B	C	D
1	**Speedometer**		**Pointer**	
2	Start	0	Value	70
3	Initial	15	Pointer	1
4	Middle	45	End	129
5	End	40		
6	Max	100		

Fig. 2.39

Note that the Doughnut series (Data 1) has 5 data points and the Pie series (Data 2) has 3 data points.

1. Select Insert > Charts > Combo.

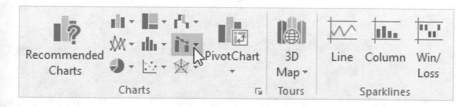

Fig. 2.40

2. Select the option "Create Custom Combo Chart."

Fig. 2.41

The Insert Chart dialog box pops up.

3. For the Doughnut series, select the chart type "Doughnut." For the Pie series, select the Pie chart type "Pie."
4. Check the box for the secondary axis for the Pie series. Click OK.

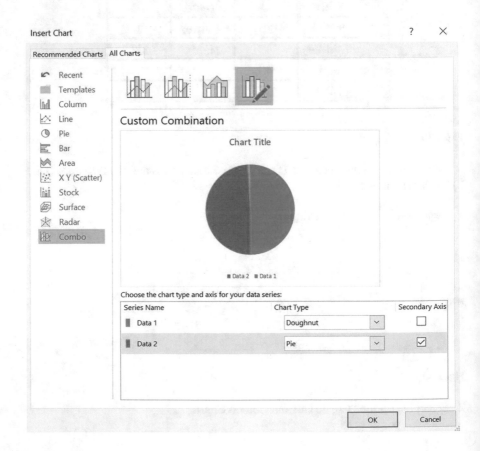

Fig. 2.42

5. Omit the legend and the chart title.
6. Select the chart and go to Format > Current Selection, then select the "Data 2 series" for the Pie series.

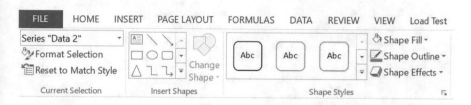

Fig. 2.43

7. Go to Format > Current Selection > Format Selection. Then set the "Angle of first slice" to 270°.

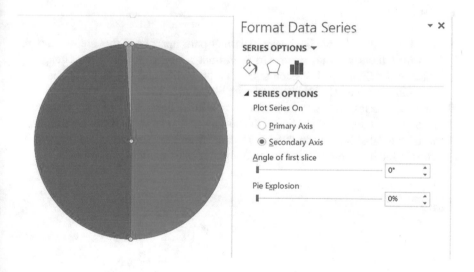

Fig. 2.44

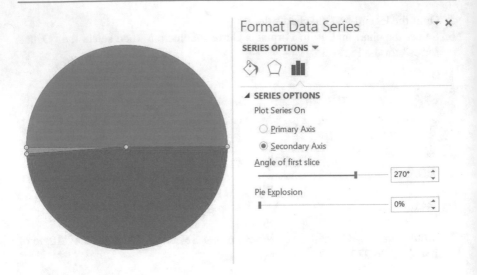

Fig. 2.45

8. The 1% slice of the pie is chosen. All other parts apart from that are made white by filling them with no color. Go to Format > Shape Styles. Then change the color of the Shape Fill of each slice. The 50% slice and the 49% slice is made "No Fill," but the 1% slice is filled with black.

9. Now, similar tasks have to be performed for the doughnut chart. Firstly, the chart is rotated to take the 50% slice at the bottom so that later on, it can be eliminated with "No Fill." Select the "Series Data 1" from the "Current Selection" of the Format section. This will make the doughnut chart available for rotation.

Fig. 2.46

10. Go to Format > Current Selection > Format Selection. Then set the "Angle of first slice" to 270°. The chart will be as follows.

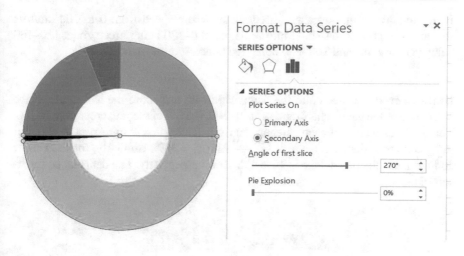

Fig. 2.47

11. The final task is to omit the 50% slice from the doughnut chart to make it a gauge chart. Go to Format > Shape Styles. Then change the color of the Shape Fill of the 50% slice to "No fill."

Result:

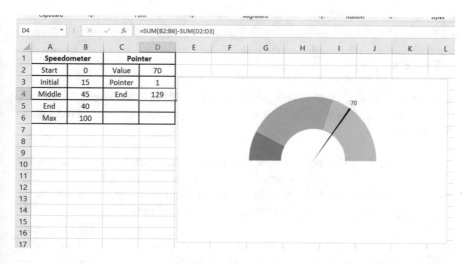

Fig. 2.48

12. Select the chart and go to Format > Current Selection > Chart Area. From the group named Shape Styles, set the Shape Fill = "No fill" and the Shape Outline = "No Outline."

Exercise 2.4: Design a gauge speedometer where speed from 0 to 200 km/h is shown. The speed is classified into three parts: 0–120 being green zone, 120–160 being yellow zone, and 160–200 being red zone. Add the marker.

Hint: Convert the speed into the percentage with respect to the half-circle. If the half-circle of the gauge chart represents 0–200, the full circle must represent 0–400. Thus, the percentage of each category has to be calculated with respect to 400. For the green zone, the percentage is $\dfrac{120-0}{400}*100\% = 30\%$. Similarly, for the yellow and red zone, the percentage is 10% and 10%, respectively. The data can be represented as:

Data 1	Data 2
30%	50%
10%	49%
10%	1%
50%	

Fig. 2.49

2.5   Bar Chart

A *bar chart* may be comparable to a column chart, but the columns are placed horizontally. A bar chart is used for large text labels. There are bar charts of the following categories:

(i)   *Clustered Bar*: This chart is used to compare values based on some categories.
(ii) *Stacked Bar*: This chart compares the whole across categories.
(iii) *100% Stacked Bar*: This chart shows the contribution of the elements based on their percentage. It is suitable for a large amount of data.

Team	Members
Circuitry	36
Software	18
Communication	25
Mechanical	12
Management	9

To form a 2D-clustered bar chart for the above data, execute this process:

1. Select the cells A1:B6.

	A	B	C
1	**Team**	**Members**	
2	Circuitry	36	
3	Software	18	
4	Communication	25	
5	Mechanical	12	
6	Management	9	
7			

Fig. 2.50

2. Go to Insert > Charts > Column > Clustered bar.

Fig. 2.51

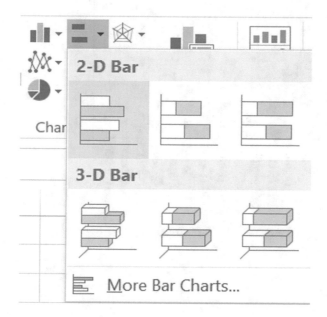

Fig. 2.52

Result:

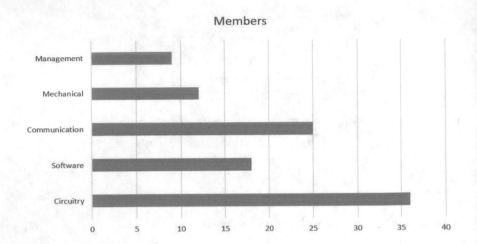

Fig. 2.53

Exercise 2.5: Make a Bar chart using the information of this column chart.

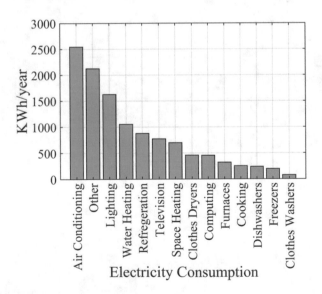

Fig. 2.54

2.6   Area Chart

An *area chart* is a modified version of a line chart with the area under the line filled with colors. Both show the trend with respect to time. Area charts are useful to show information with several time series data. They can also show the share of each dataset to the whole. The 2D and 3D area charts of the following categories:

1. *Area Chart*: This chart is used for two types of data, an overall dataset, and a subset. There is overlap in the colors if the data values are overlapping.
2. *Stacked Area Chart*: This chart is used in most of the cases, as the colors are distinguished in this chart.
3. *100% Stacked Area Chart*: This chart amounts all the values in the Y-axis to 100%.

Month	Product-A	Product-B	Product-C	Product-D
Jan	1289	934	393	728
Mar	1358	239	590	1209
May	934	738	853	984
Jul	1532	794	542	1002
Sept	1340	987	432	803
Nov	1193	881	928	824

To generate a stacked area chart for the given dataset, follow this process:

1. Select the cells A1:D7.

	A	B	C	D	E	F
1	Month	Product A	Product B	Product C	Product D	
2	Jan	1289	934	393	728	
3	Mar	1358	239	590	1209	
4	May	934	738	853	984	
5	Jul	1532	794	542	1002	
6	Sept	1340	987	432	803	
7	Nov	1193	881	928	824	
8						
9						

Fig. 2.55

2. Go to Insert > Charts > Area Chart.

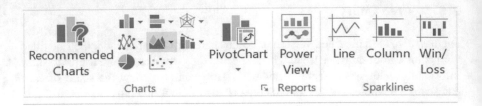

Fig. 2.56

3. Click Area Chart.

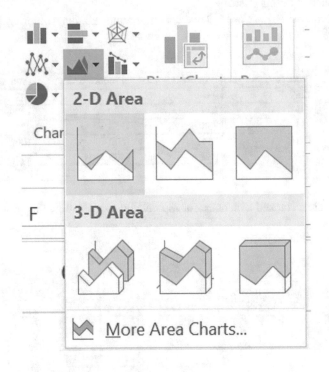

Fig. 2.57

Result. In this example, some areas overlap.

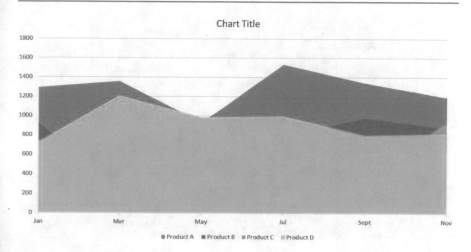

Fig. 2.58

From Fig. 2.59, you can find the corresponding line chart to clearly see the reason behind the overlapping.

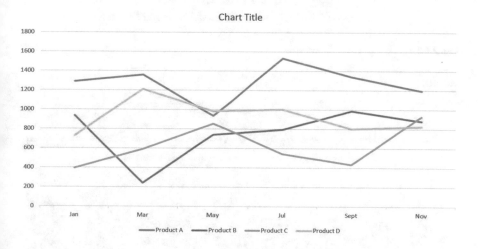

Fig. 2.59

4. Change the chart's subtype to Stacked Area from the "chart type" option.

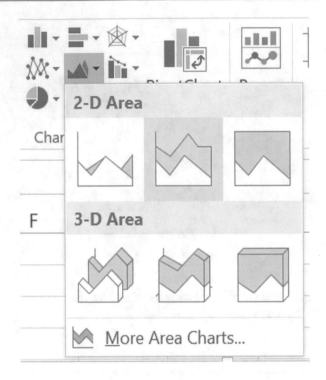

Fig. 2.60

Result:

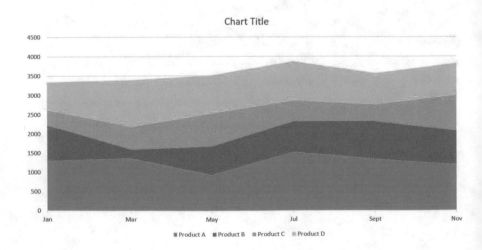

Fig. 2.61

Exercise 2.6: Plot this area chart. Change the color and style of the chart.

Region	Quantity 1	Quantity 2	Quantity 3
1	23	32	53
2	56	76	46
3	45	35	24
4	5	98	85

2.7   Scatter Chart

A *scatter chart* (XY chart) can be used to display scientific data involving interdependent variables. It is used to determine a relationship between variables X and Y. The scatter charts are represented in the following ways:

1. *Scatter*: This chart is used to compare at least two pairs of data to depict the relationship between two sets of values.
2. *Scatter with straight lines and/or markers*: Markers are usually used when there are a few data to plot. For a large number of data, markers will overlap, and the points will overlap. Thus, it is better not to use markers for a large amount of data.
3. *Scatter with smooth lines and/or markers*: The points in the chart are added using a smooth line. The marker is usually used for a small amount of data. Moreover, this chart is normally used where the relation between X and Y is determined using a formula.

Time (s)	Resistance (Ohm)	Voltage (V)	Power (W)
0	0	9	0
1	10	27	3
2	20	33	7
3	30	45	9
4	40	66	15
5	50	69	24
6	60	74	32
7	70	93	38
8	80	86	39
9	90	98	44
10	100	112	58

Follow the step hereunder to generate a scatter plot from the above Table.

1. Select the cells A1:D10.

	A	B	C	D	E
1	Time	Resistance	Voltage	Power	
2	0	0	0	0	
3	1	10	27	3	
4	2	20	33	7	
5	3	30	45	9	
6	4	40	66	15	
7	5	50	69	24	
8	6	60	74	32	
9	7	70	93	38	
10	8	80	86	39	
11	9	90	98	44	
12	10	100	112	58	
13					

Fig. 2.62

2. Go to Insert > Charts > Scatter.

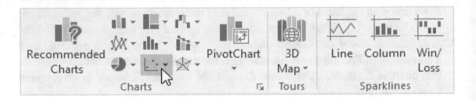

Fig. 2.63

3. At first, click on Scatter without lines.

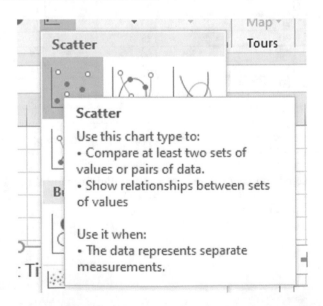

Fig. 2.64

The graph will be like Fig. 2.65.

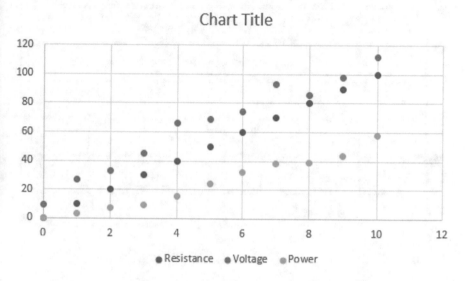

Fig. 2.65

4. Then, click on Scatter with lines

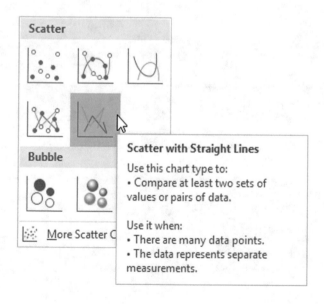

Fig. 2.66

The graph appears as follows:

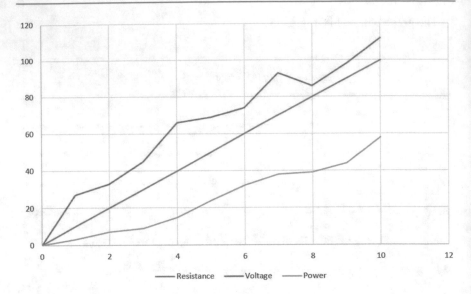

Fig. 2.67

5. Now click on the subtype "Scatter with Smooth Lines" to see the following curve.

Result:

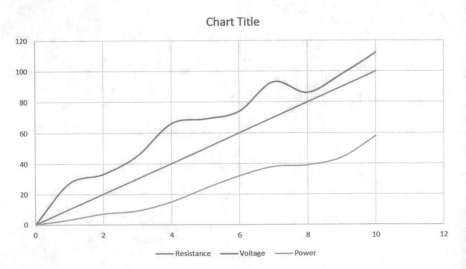

Fig. 2.68

N.B. A title is added for both the horizontal and the vertical axis. In a scatter plot, the horizontal axis is the value axis, with more options for scaling the axis.

2.7.1   Switching Between X and Y-Axis

Consider a scatter chart as Fig. 2.69 for the following data. Now we want to swap the X- and the Y-axis.

X	Y
0	0
2	33
4	56
6	23
8	44
10	50
12	74
14	21
16	64
18	11
20	98

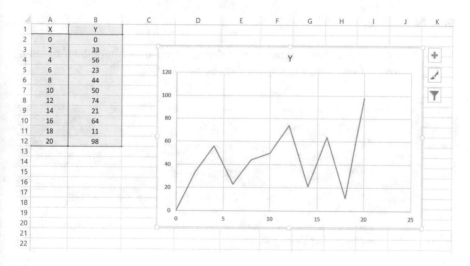

Fig. 2.69

To switch the X and Y axes, do the following:

1. Right-click on the scatter chart. Then click on Select Data from the context menu.

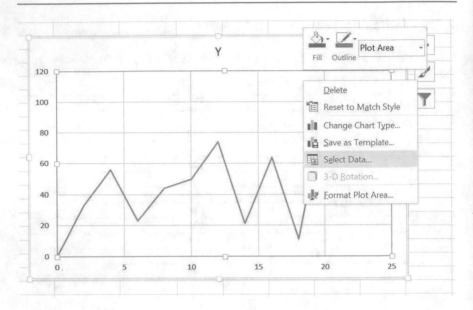

Fig. 2.70

2. On the dialog box named "Select Data Source," click to highlight the Y column. Then click on Edit in the Legend Entries (Series) section.

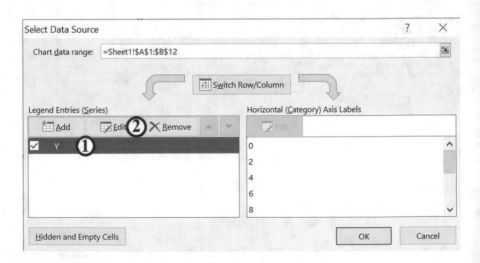

Fig. 2.71

3. Now the Edit Series dialog box pops up. Swap the values of Series X and Series Y. Click OK for both boxes.

	A	B	C	D	E	F	G	H	I	J	K	L
2	0	0										
3	2	33										
4	4	56	Edit Series								?	X
5	6	23	Series name:									
6	8	44	=Sheet1!$B$1		= Y							
7	10	50	Series X values:									
8	12	74	=Sheet1!$B$2:$B$12		= 0, 33, 56, 23,...							
9	14	21	Series Y values:									
10	16	64	=Sheet1!$A$2:$A$12		= 0, 2, 4, 6, 8,...							
11	18	11										
12	20	98							OK		Cancel	
13												

Fig. 2.72

Now the X- and Y-axis are switched in the scatter chart.

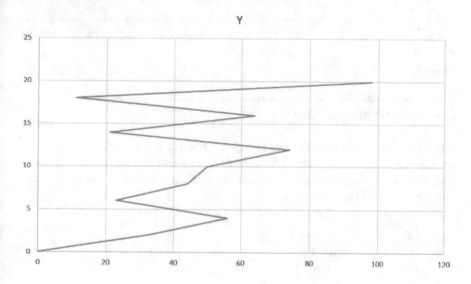

Fig. 2.73

Exercise 2.7: From the given data, take the data of time and voltage and plot the scatter plot. Try to switch the axis data and plot. Also, try to add the trendline (mentioned later in Sect. 5.4.2). The original diagram with the trendline should look like this.

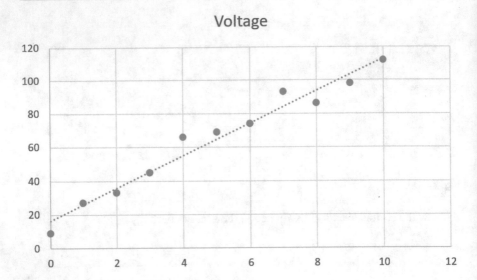

Fig. 2.74

2.8   Combination Chart

A *combination chart* or combo chart combining two or more types of charts. There are three types of combo charts, but the most general case can be visited by customizing the charts. To generate a combination chart for the data in the table below, execute the steps that follow.

Month	Working days	Revenue
Jan	13	$1039
Feb	15	$968
Mar	22	$2983
Apr	19	$5976
May	28	$2409
Jun	23	$845
Jul	14	$2309
Aug	11	$1902
Sep	20	$3924
Oct	24	$955
Nov	15	$3405
Dec	26	$2375

1. Select the range A1:C13.

	A	B	C	D
1	Month	WorkingDays	Revenue	
2	Jan	13	$1,039	
3	Feb	15	$968	
4	Mar	22	$2,983	
5	Apr	19	$5,976	
6	May	28	$2,409	
7	Jun	23	$845	
8	Jul	14	$2,309	
9	Aug	11	$1,902	
10	Sep	20	$3,924	
11	Oct	24	$955	
12	Nov	15	$3,405	
13	Dec	26	$2,375	
14				

Fig. 2.75

2. Go to Insert > Charts > Combo.

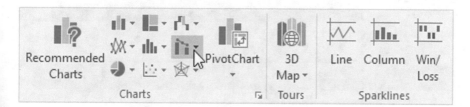

Fig. 2.76

3. Select "Create Custom Combo Chart."

Fig. 2.77

The Insert Chart dialog box appears. To make a Clustered Column-Line on Secondary Axis chart, do as follows:

4. Select the Clustered Column chart type for the Working Days series.
5. Select the type Line chart for the Revenue series.
6. Under the secondary axis, plot the Revenue series.

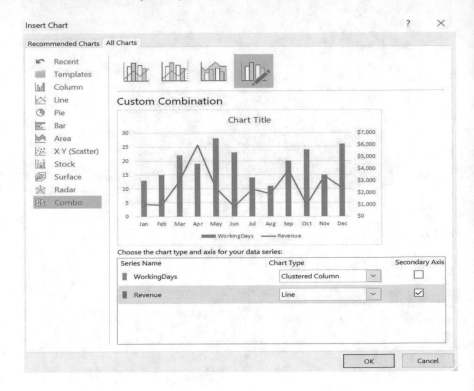

Fig. 2.78

7. Click OK.

Result:

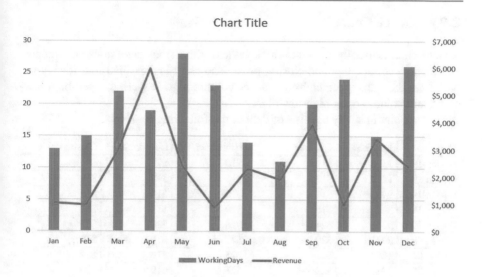

Fig. 2.79

Exercise 2.8: Using the table provided in this section, plot the combinational chart combining an area chart and a scatter chart in the secondary axis. The graph should look like this:

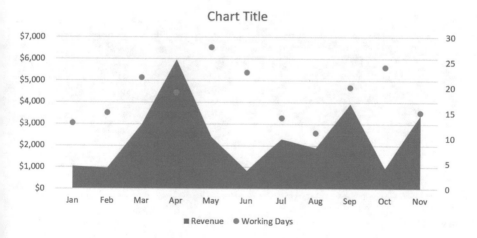

Fig. 2.80

2.9   Stock Chart

A stock chart is usually related to stock values, with a stock date, stock volume, low
and high price, opening and closing price. The chart is classified into four types,
based on the data type: high-low-close, volume-high-low-close, open-high-low-
close, and volume-open-high-low-close.

A stock chart usually consists of data of the following format:

Date	Volume	Opening price	High price	Low price	Closing price
01-Jan-19	1328	$11	$10	$7	$8
02-Jan-19	1283	$13	$15	$12	$14
03-Jan-19	943	$14	$16	$15	$15
04-Jan-19	857	$16	$19	$13	$15
05-Jan-19	923	$24	$28	$20	$25
06-Jan-19	2321	$20	$23	$17	$19
07 Jan 19	1242	$26	$25	$16	$22
08-Jan-19	1983	$22	$21	$19	$20
09-Jan-19	1343	$18	$17	$15	$16
10-Jan-19	2931	$20	$22	$18	$21

To draw a stock chart, follow the procedures as follows:

1. Select A1:F11.

	A	B	C	D	E	F	G
1	Date	Volume	Open Price	High Price	Low Price	Closing Price	
2	01-Jan-19	1328	$11	$10	$7	$8	
3	02-Jan-19	1283	$13	$15	$12	$14	
4	03-Jan-19	943	$14	$16	$15	$15	
5	04-Jan-19	857	$16	$19	$13	$15	
6	05-Jan-19	923	$24	$28	$20	$25	
7	06-Jan-19	2321	$20	$23	$17	$19	
8	07 Jan 19	1242	$26	$25	$16	$22	
9	08-Jan-19	1983	$22	$21	$19	$20	
10	09-Jan-19	1343	$18	$17	$15	$16	
11	10-Jan-19	2931	$20	$22	$18	$21	
12							

Fig. 2.81

2. Click "Insert Stock, Surface or Radar Chart."

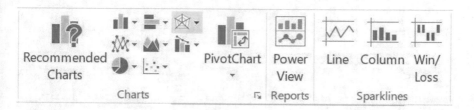

Fig. 2.82

3. To draw a stock chart of the type volume-open-high-low-close, select the following.

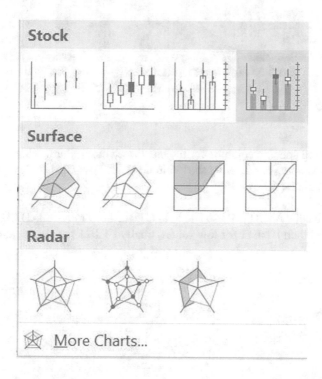

Fig. 2.83

The result will be as follows:

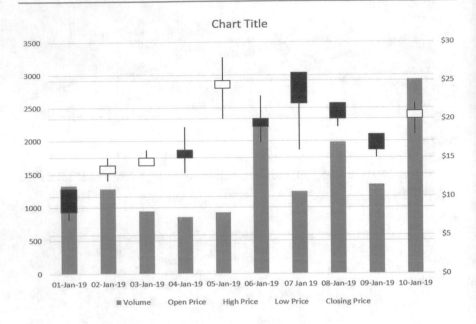

Fig. 2.84

By selecting specific columns for the mentioned stock chart, other charts can be drawn. For example, for drawing the "Volume-High-Low-Close" curve, the following steps should be followed:

1. Select the cells A1:B11. Press and hold CTRL. Select the cells D1:D11 for high values and then E1:E11 for low values, finally F1:F11 for closed values, and let go CTRL.

	A	B	C	D	E	F	G
1	Date	Volume	Open Price	High Price	Low Price	Closing Price	
2	01-Jan-19	1328	$11	$10	$7	$8	
3	02-Jan-19	1283	$13	$15	$12	$14	
4	03-Jan-19	943	$14	$16	$15	$15	
5	04-Jan-19	857	$16	$19	$13	$15	
6	05-Jan-19	923	$24	$28	$20	$25	
7	06-Jan-19	2321	$20	$23	$17	$19	
8	07 Jan 19	1242	$26	$25	$16	$22	
9	08-Jan-19	1983	$22	$21	$19	$20	
10	09-Jan-19	1343	$18	$17	$15	$16	
11	10-Jan-19	2931	$20	$22	$18	$21	
12							

Fig. 2.85

2. Click "Insert Stock, Surface or Radar Chart," select "Volume-High-Low-Close Stock Chart."

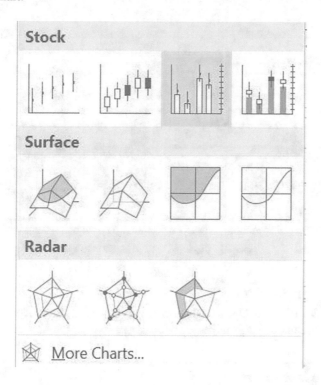

Fig. 2.86

3. The result will be as follows:

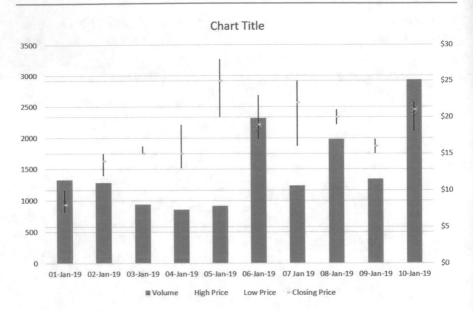

Fig. 2.87

Exercise 2.9: Plot the Open-High-Low-Close and High-Low-Close stock chart from the above data.

2.10   Surface Chart

Surface charts are mostly used to determine the optimum data between two sets. The chart is in the form of a topographic map with patterns and colors to represent the change in data.

1. 3-D Surface: It shows a 3-dimensional representation of a column chart wrapped with a sheet, where the height varies according to the frequency. A clear idea about the data cannot be determined, as such a chart shows the differences between two values in a continuous curve.
2. Wireframe 3-D Surface: It is similar to a 3-D surface, but instead of colors, it is displayed with the lines. It is usually used to show the trend of values and the frequency of appearance of a certain data type.
3. Contour: Contour is the top view of the 3-D surface chart. It is a 2-D map similar to a topography, where the frequency and data range is represented with colors.
4. Wireframe 3-D Contour: It resembles the contour, but instead of colors, it is drawn with only the curves or lines. It creates the boundary around the data of a similar type.

For the following dataset, the curve can be determined as follows:

Coffee	January to March	April to June	July to September	October to December
Latte	5323	3244	4535	5757
Cappuccino	2321	2112	1212	6456
Espresso	6256	1312	4557	7531
Mochaccino	5366	7653	1231	1334
Milkshakes	5453	3134	8243	1329

1. Select A1:E6. Go to the ribbon and click "3-D Surface Chart."

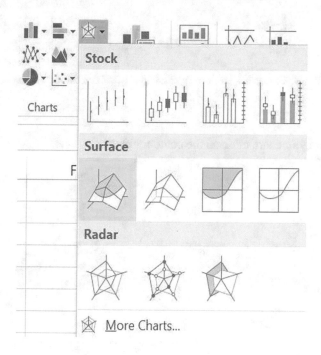

Fig. 2.88

2. This will result in the following diagram.

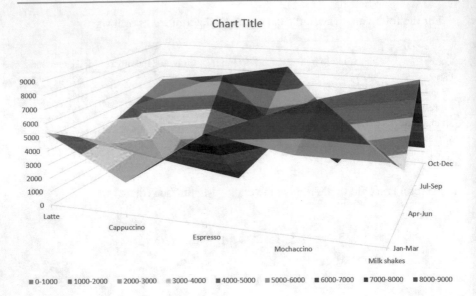

Chart Title

Fig. 2.89

For the contour chart, click on the contour chart.

Result:

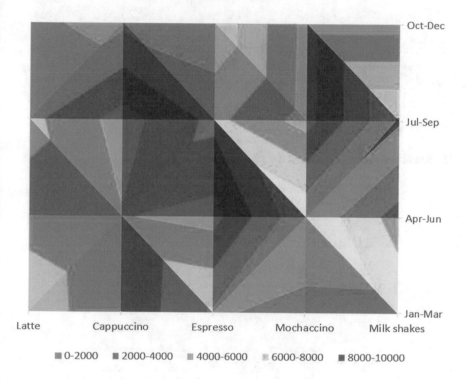

Chart Title

Fig. 2.90

Exercise 2.10: Plot the Wireframe 3-D Surface chart and the wireframe contour from the above data and compare with the given 3-D Surface and contour chart, respectively.

2.11   Radar Chart

A radar chart is also known as a polar chart, web chart, spider chart, or star plots to represent multivariate data to plot a set of data over multiple common variables. The radar chart is usually of general type, with markers or areas filled with colors.

To draw the radar chart for the following data, follow the instructions:

Seasons	12 pm to 6 pm	6 pm to 12 am	12 am to 6 am	6 am to 12 pm
Summer	60	20	30	50
Fall	20	50	45	35
Winter	10	40	50	20

1. Select A1:E4 after copying the data in an Excel sheet.

	A	B	C	D	E	F
1	Seasons	12pm-6pm	6pm-12am	12am-6am	6am-12pm	
2	Summer	60	20	30	50	
3	Fall	20	50	45	35	
4	Winter	10	40	50	20	
5						

Fig. 2.91

2. Select the Radar chart from the ribbon.

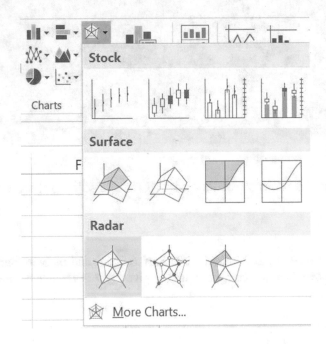

Fig. 2.92

3. After clicking the Radar chart, the following figure will appear.

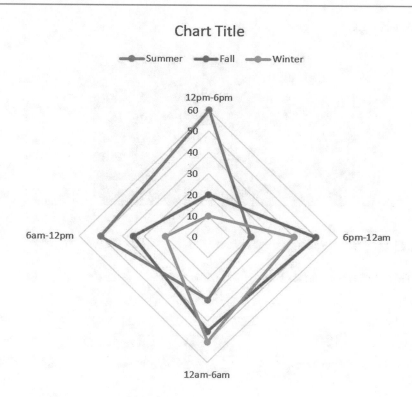

Fig. 2.93

Exercise 2.11: Plot the Radar chart with a marker and filled radar chart from the above data.

2.12   Gantt Chart

There is no chart type that readily provides a Gantt chart in Excel. But a Gantt chart can be produced upon customizing a stacked bar chart. Consider the following data:

Item	Date	Days
Fundraising	01-Sep-20	28
Purchasing the component	01-Oct-20	13
Designing the hardware	15-Oct-20	14
Integration	28-Oct-20	14
Coding	12-Nov-20	17
Simulation	01-Dec-20	18
Completion	20-Dec-20	20

For creating a Gantt chart, follow these steps:

1. Select A1:C8.

	A	B	C	D
1		Date	Days	
2	Fundraising	01-Sep-20	28	
3	Purchasing the component	01-Oct-20	13	
4	Designing the hardware	15-Oct-20	14	
5	Integration	28-Oct-20	14	
6	Coding	12-Nov-20	17	
7	Simulation	01-Dec-20	18	
8	Completion	20-Dec-20	20	
9				

Fig. 2.94

2. Go to Insert > Charts > Bar.

Fig. 2.95

3. Click Stacked Bar.

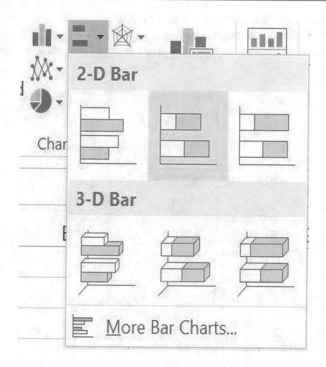

Fig. 2.96

Result:

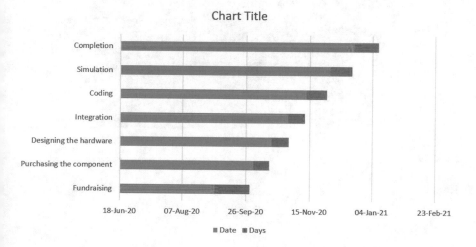

Fig. 2.97

4. Click on Chart Title and insert a suitable title, such as "Project Timeline."

5. At the bottom, click the legend and press Delete.
6. The tasks (Fundraising, Purchasing the component, etc.) are in the reverse order.
    Right-click on the tasks. Then go to Format Axis > Categories in reverse order.

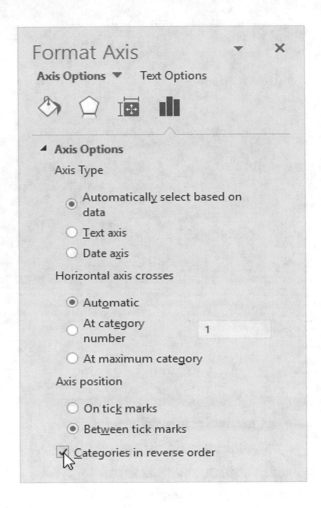

Fig. 2.98

Result:

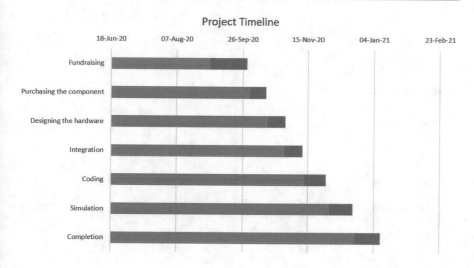

Fig. 2.99

7. On the blue bars, right-click > Format Data Series > Fill & Line icon > Fill > No fill.

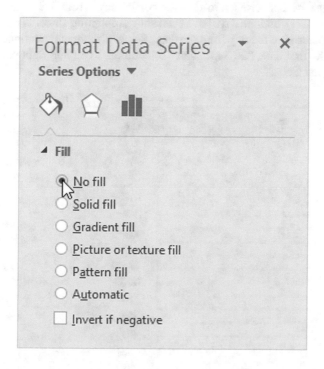

Fig. 2.100
The graph becomes:

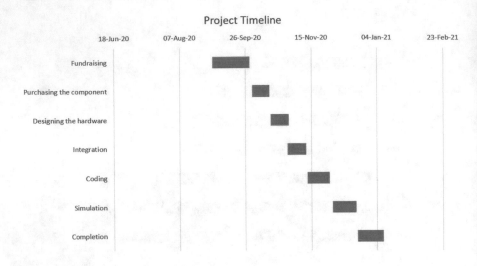

Fig. 2.101

8. It is seen that the data are in the middle, so the maximum and minimum bound of the date is needed to be set. Date and time are stored in MS Excel as integer numbers according to the number of days passed since January 1, 1900. So, the dates must be converted into the number of days from 1 January 1900 (for Windows User), and the days have been started counting since. There are several ways to determine the day from the date. Some of them are mentioned below:

(i) For the first date, "20 Sept 2020," select the B2 cell and right-click to select "Format Cells."

Fig. 2.102

(ii) From the dialog box named "Format Cells," select "General" and note down the integer number that appears in the box, for this case, 44075, which will be our minimum bound for the date. Click "OK."

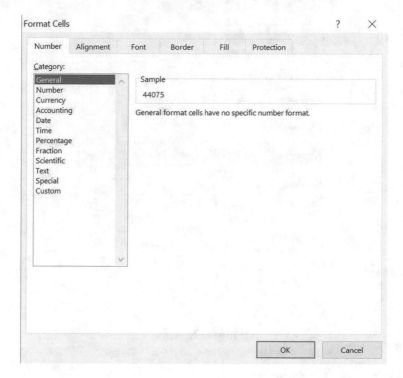

Fig. 2.103

(iii) Similarly, determine the integer number for the date "20 Dec 2020," which will turn out to be 44185. But as the duration is of 20 more days, that means 20 will be added with 44185, which will be 44205. Thus, the maximum bound for displaying the graph will be 44205.

9. As the maximum and minimum bounds are determined, right-click on the date axis and go to "Format Axis." Type 44075 as the minimum and 44205 as the maximum bounds.

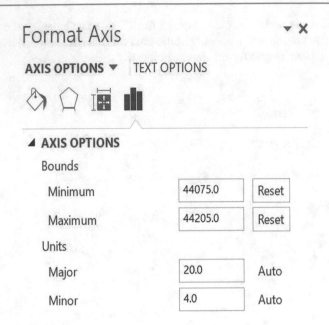

Fig. 2.104

Result:

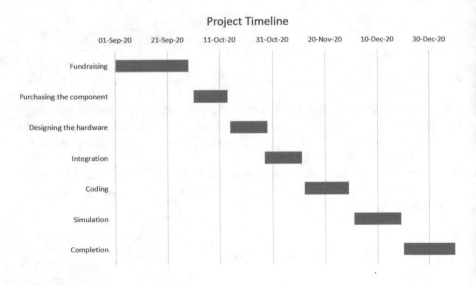

Fig. 2.105

Exercise 2.12: Build the Gantt chart from the following data:

Task list	Date	Days
Task one	01-Jan-15	35
Task two	15-Feb-15	14
Task three	01-Mar-15	14
Task four	15-Apr-15	44
Task five	30-May-15	12
Task six	15-Jun-15	38
Task seven	01-Aug-15	30
Task eight	15-Sep-15	21
Task nine	15-Oct-15	58
Task ten	20-Dec-15	8

2.13   Pareto Chart

As per the Pareto principle, for most events, about 80% of effects originate from 20% causes. The Pareto principle is one of the most important quality tools used in industries for quality inspection. Consider the following data from a beverage company:

Reasons of complaints	Number of customers
Overpriced	334
Quantity is low	57
Taste is bad	31
Too sweet	9
Not available in bottles	104
Poor marketing	65

The Pareto Chart is basically a combinational chart with clustered column charts of the complaints and a line chart of the cumulative percentage of the data. Thus, to make the cumulative percentage, sort the number of customers in descending order and add those sequentially. Finally, convert the values into a percentage so that the total number of customers corresponds to 100%. The table should look like this.

Reasons of complaints	Number of customers	Reasons of complaints	Sort	Cumulative summation (x_i)	Cumulative percentage = $(x_i/600)$ * 100%
Overpriced	334	Overpriced	334	334	55.67%
Quantity is low	57	Not available in bottles	104	438	73%
Taste is bad	31	Poor marketing	65	503	83.83%

(Continued)

Reasons of complaints	Number of customers	Reasons of complaints	Sort	Cumulative summation (x_i)	Cumulative percentage = $(x_i/600)$ * 100%
Too sweet	9	Quantity is low	57	560	93.33%
Not available in bottles	104	Taste is bad	31	591	98.5%
Poor marketing	65	Too sweet	9	600	100%

To form a Pareto chart, follow this process.

1. Copy the last four columns of the above table in Excel.

	A	B	C	D	E
1	Reasons of complaints	Sort	Cumulative summation (x_i)	Cumulative percentage= $(x_i/600)*100\%$	
2	Overpriced	334	334	55.67%	
3	Not available in bottles	104	438	73%	
4	Poor marketing	65	503	83.83%	
5	Quantity is low	57	560	93.33%	
6	Taste is bad	31	591	98.50%	
7	Too sweet	9	600	100%	
8					

Fig. 2.106

2. Select A1:B7, then press CTRL and select D1:D7. Go to Insert > Charts > Insert Combo Chart.

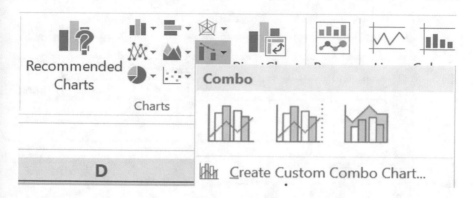

Fig. 2.107

3. Select "Create Custom Combo Chart."
4. Choose "Clustered Column" for the "Sort" field and "Line Chart" for the "Cumulative Percentage" field. Put it at the "Secondary Axis."

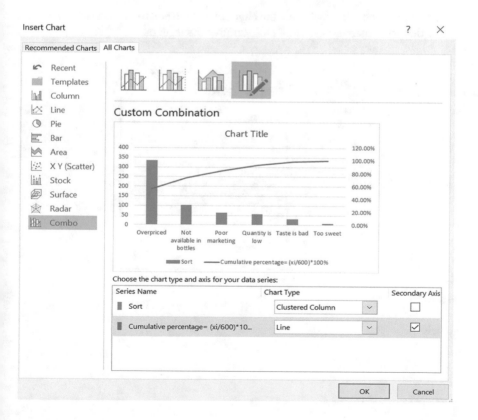

Fig. 2.108

5. Click "OK." The graph should look like this:

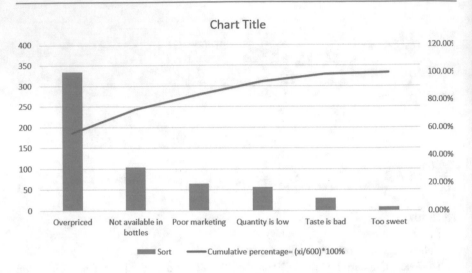

Fig. 2.109

6. Clicking on the + button beside the chart, check the box for Data Labels. Rename the title to "Customer Complaints." Remove the legends.

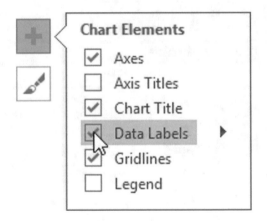

Fig. 2.110

Result:

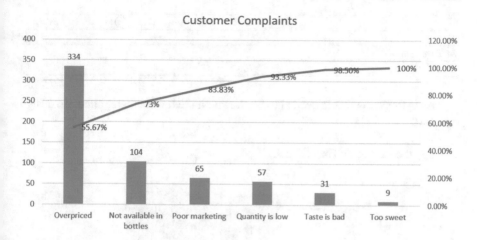

Fig. 2.111

Exercise 2.13: Draw the Pareto Chart from the following data:

Comments on teacher	Number of students
Poor teaching style	35
Lacks practicality	46
Bad grading	22
Less friendly	7
Delays in returning grade sheets	13
Follows a single textbook	27

2.14   Thermometer Chart

A thermometer chart represents what fraction of a goal has been achieved. Consider the following data, which indicates the charging status of a battery according to the hour passed.

Hour	Increase in charging status (%)
1	7
2	8
3	5
4	7
5	10
6	13
7	16
8	8
9	

(Continued)

Hour	Increase in charging status (%)
10	
Achievement	74
Goal	100

74% is the value that is needed to be shown on the thermometer chart. To design the chart:

1. Copy the data in an Excel sheet.

	A	B
1	Hour	Increase in Charging Status (%)
2	1	7
3	2	8
4	3	5
5	4	7
6	5	10
7	6	13
8	7	16
9	8	8
10	9	
11	10	
12	Achievement	74
13	Goal	100
14		

Fig. 2.112

2. Select only cell B12, which indicates the percentage.
3. Go to Insert > Charts > Column.

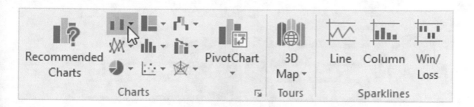

Fig. 2.113

4. Click Clustered Column.

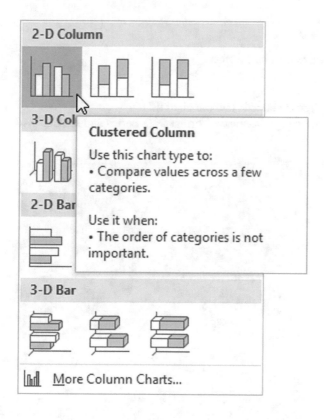

Fig. 2.114

Result:

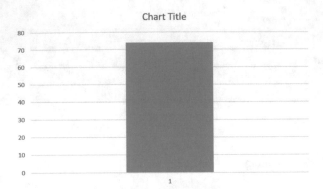

Fig. 2.115

To customize the chart further:

5. Omit the horizontal axis and the chart title.
6. On the blue bar, right-click > Format Data Series and set Gap Width = 0%.

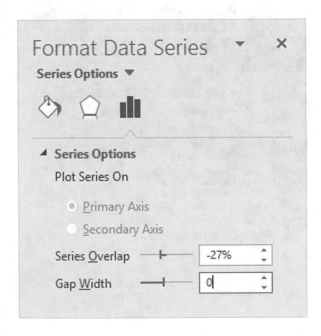

Fig. 2.116

7. Change the chart width.

8. Right-click on the percentages on the chart. Then go to Format Axis and set the minimum bound = 0 and the maximum bound = 1. Choose "Outside" as the major tick mark.

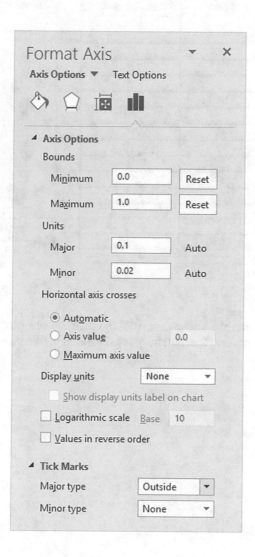

Fig. 2.117

Result:

	A	B	C	D
	Hour	**Increase in Charging Status (%)**		
1				
2	1	7		
3	2	8		
4	3	5		
5	4	7		
6	5	10		
7	6	13		
8	7	16		
9	8	8		
10	9			
11	10			
12	**Achievement**	74		
13	**Goal**	100		
14				

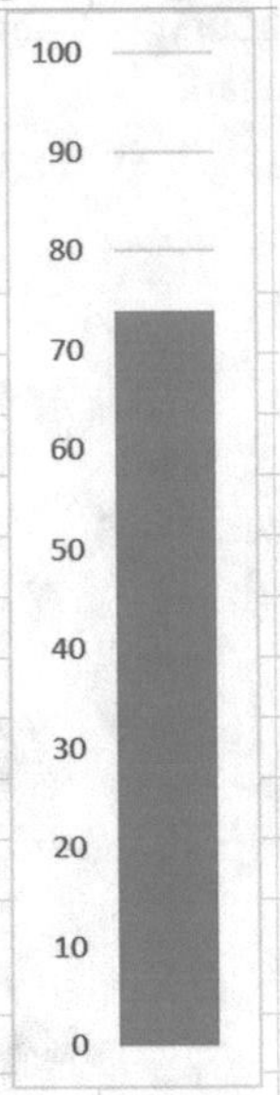

Fig. 2.118

Exercise 2.14: From the above data, plot the thermometer plot of the data if the 15% charge increases on Hour 9. The graph should look like this.

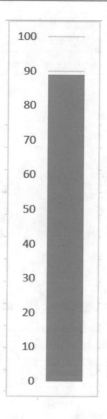

Fig. 2.119

2.15   Sparklines

Sparklines in *Excel* are graphs that fit in one cell and give you the information or trend about the data. The sparklines are usually of three types:

2.15.1 Line

It provides or displays the trend or change of data of a row in Excel. For the following data, the line can be drawn by executing the following steps:

1	2	3	4	5	6
3	9	6	12	18	15
35	30	25	20	15	10

1. Copy the data in Excel. Select blank cells G1:G3 as the sparkline location.

	A	B	C	D	E	F	G
1	1	2	3	4	5	6	
2	3	9	6	12	18	15	
3	35	30	25	20	15	10	
4							

Fig. 2.120

2. Go to Insert > Sparkline > Line.

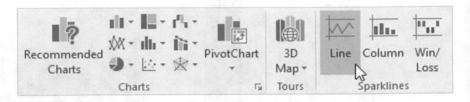

Fig. 2.121

3. Click the Data Range box and type A1:F3 to select the existing data.

Fig. 2.122

4. Click OK.

Result:

	A	B	C	D	E	F	G
1	1	2	3	4	5	6	
2	3	9	6	12	18	15	
3	35	30	25	20	15	10	
4							

Fig. 2.123

2.15.2 Column

Similar to the line chart, it shows the trend but in a column chart in cells. For the given data, go to "Sparkline Tool" at the ribbon and click on Design. Select "Column" to change the representation from line to column. The style of the sparkline, along with the color, can be changed from this ribbon.

Fig. 2.124

Result:

	A	B	C	D	E	F	G
1	1	2	3	4	5	6	
2	3	9	6	12	18	15	
3	35	30	25	20	15	10	

Fig. 2.125

2.15.3 Win/Loss

A win-loss chart can be determined if some number occurs to be negative. For example, if we consider the first row of the following data to be the selling price and the second row to be the buying price, the difference, if positive, will indicate profit or win. Negative will denote loss. The difference is shown in the third row.

Selling price	$5	$16	$5	$13	$7
Buying price	$3	$18	$10	$10	$4
Selling price − buying price = profit or loss	$2	−$2	−$5	$3	$3

To draw the Win/Loss Sparkline, do the following:

1. Copy the above data in Excel. Select A1:E3 and right-click on the data. Then, go to "Format Cells" and delete the "$" sign so that the cells can take the negative values.

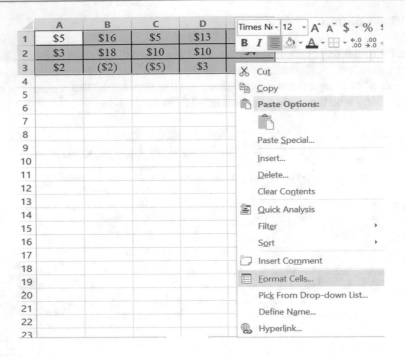

Fig. 2.126

2. After the "Format Cells" dialog box appears, choose category "General" and press OK.

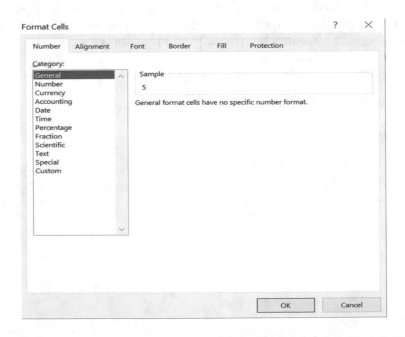

Fig. 2.127

3. The data appears as follows.

	A	B	C	D	E
1	5	16	5	13	7
2	3	18	10	10	4
3	2	-2	-5	3	3

Fig. 2.128

4. Select blank cell F3 as the sparkline location. Go to Insert > Sparkline group > Win/Loss.

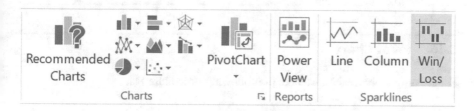

Fig. 2.129

5. In the following dialog box, type A3:E3 to indicate which data it must take to create the sparkline.

Create Sparklines   ?   ×

Choose the data that you want

Data Range: A3:E3

Choose where you want the sparklines to be placed

Location Range: $F$3

OK   Cancel

Fig. 2.130

6. Click OK.

Result:

	A	B	C	D	E	F
1	5	16	5	13	7	
2	3	18	10	10	4	
3	2	-2	-5	3	3	▪▪▪▪▪

Fig. 2.131

Exercise 2.15: Determine the line and column sparkline for the following data. Also, determine their differences (value of the first row – value of the second row) and determine the win/loss sparkline.

23	65	31	11	8	27	72
3	19	33	56	29	16	5

2.16 Pivot Chart

For creating a pivot table, execute the following procedure [1].

1. In the data set, Select any single cell.
2. Go to Insert > Tables > PivotTable. The same output will be found from Charts > PivotChart > PivotChart & PivotTable.

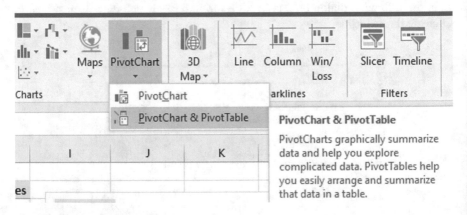

Fig. 2.132

Either way, the dialog box in Fig. 2.133 appears. Excel selects the data automatically. A pivot table is created in a New Worksheet by default.

3. Click OK.

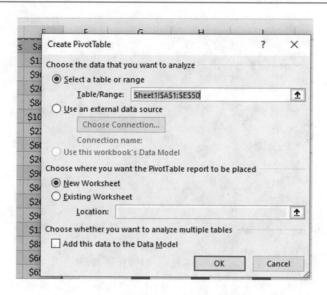

Fig. 2.133

Drag Fields

A new pane named PivotChart Fields appears. For getting the total exports of each product in a list, drag the fields to suitable areas.

1. Axis (Categories) shall contain the fields Color, Region, Months.
2. Values area shall contain the field sales.

Fig. 2.134

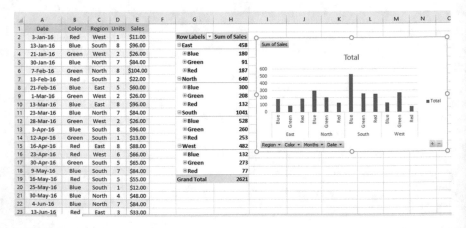

Fig. 2.135

Sorting

1. Select any cell from the column Sum of Sales.
2. Right-click on the selection and go to Sort > Sort Largest to Smallest.

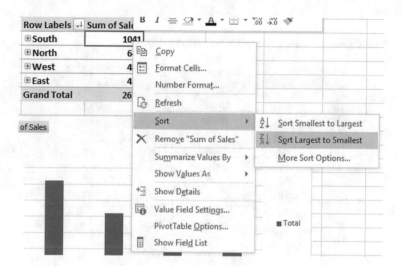

Fig. 2.136

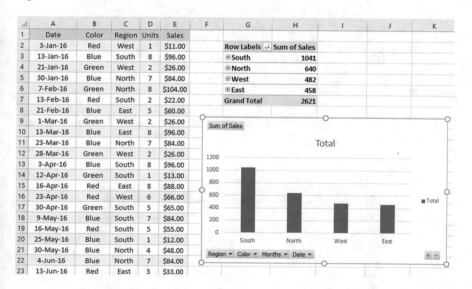

Fig. 2.137

Pivot Table

A pivot table is very useful as a data analysis tool and more flexible compared to normal or straight Tables. A normal table in Excel only shows stacks of information. A pivot table is often a derivative of a normal table that shows classified information, sums, subtotals, grand totals, mean, minima or minima, or any other information extracted from the aggregated data of the normal table.

Steps:

1. Show the data on which the work will be done.

	A	B	C	D
1	Location	Item	Price	Unit
2	North	Tomato	$12.50	13
3	North	Onion	$14.25	15
4	South	Potato	$17.35	12
5	East	Potato	$19.99	25
6	West	Cabbage	$15.50	10
7	East	Tomato	$16.00	15
8	North	Carrot	$11.75	18
9	West	Spinach	$15.00	8
10	South	Spinach	$11.25	12
11	South	Tomato	$17.35	17
12	East	Spinach	$18.00	10
13	West	Potato	$15.50	12
14	South	Onion	$13.33	19
15	South	Onion	$14.75	16
16	South	Onion	$12.99	23
17	South	Cabbage	$16.00	18
18				

Fig. 2.138

The target is to develop a table as follows where we can see which location buys which item at what cost in brief.

Sum of Price	Column Labels				
Row Labels	East	North	South	West	Grand Total
Cabbage			16	15.5	31.5
Carrot		11.75			11.75
Onion		14.25	41.07		55.32
Potato	19.99		17.35	15.5	52.84
Spinach	18		11.25	15	44.25
Tomato	16	12.5	17.35		45.85
Grand Total	53.99	38.5	103.02	46	241.51

Fig. 2.139

We will learn about the pivot table while solving this problem.

2. Select a cell on the data.
3. Insert > Pivot Table.

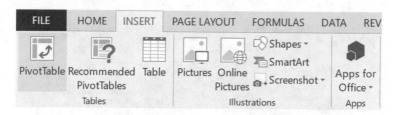

Fig. 2.140

4. Select the "Existing Worksheet" option as we are planning to save it in the same sheet. Select a cell; for G1, the location will be filled with "PIVOT!$G$1."
5. Click OK.
6. We will see the following setup.

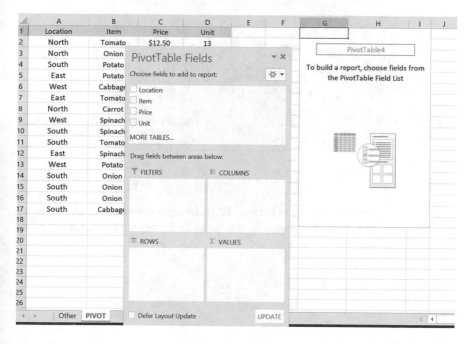

Fig. 2.141

In the G1 cell, there is an empty pivot table that requires formatting. In the pivot table field section, there are four fields (as per the table) and four areas (filter, column, rows, values) where the fields are to be distributed as per our need.

As we want to keep the name of the item as the row, the name of the location as the column, and the prices as the values, we need to drag and drop the fields in the area accordingly.

7. First, drag and drop the "item" field in the "rows" area.

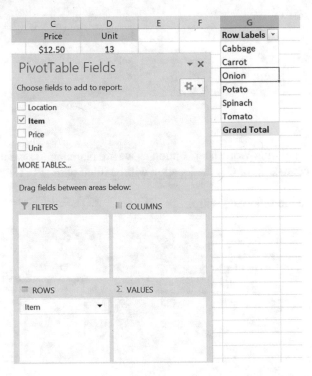

Fig. 2.142

Similarly, drag and drop the other two fields in the corresponding areas. As we drag and drop each field, we gradually see the table forming. The end result looks like this:

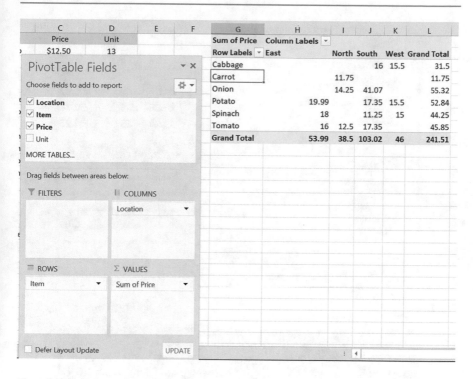

Fig. 2.143

Hence, the table is formed. Let us analyze the derived table to understand the advantage of a pivot Table.

1. From the dropdown menu in the column/row table, we can sort or label our desired column/row as per our need. For example, to sort the column in the descending (Z-A) order, we can click the dropdown as follows and select "Sort Z to A."

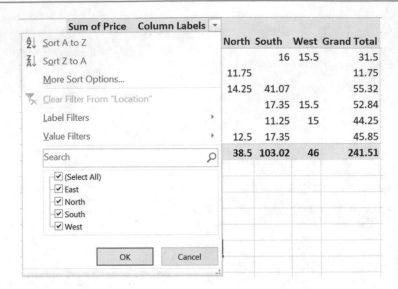

Fig. 2.144

G	H	I	J	K	L
Sum of Price	Column Labels ↓↑				
Row Labels ▾	West	South	North	East	Grand Total
Cabbage	15.5	16			31.5
Carrot			11.75		11.75
Onion		41.07	14.25		55.32
Potato	15.5	17.35		19.99	52.84
Spinach	15	11.25		18	44.25
Tomato		17.35	12.5	16	45.85
Grand Total	46	103.02	38.5	53.99	241.51

Fig. 2.145

2. If we want to determine the number of times each item has been used, we can follow the procedure below:

Drag and drop the "item" field in the "rows" area to display them as a row first. Then drop the same field in the area "values" for counting their appearance in the table.

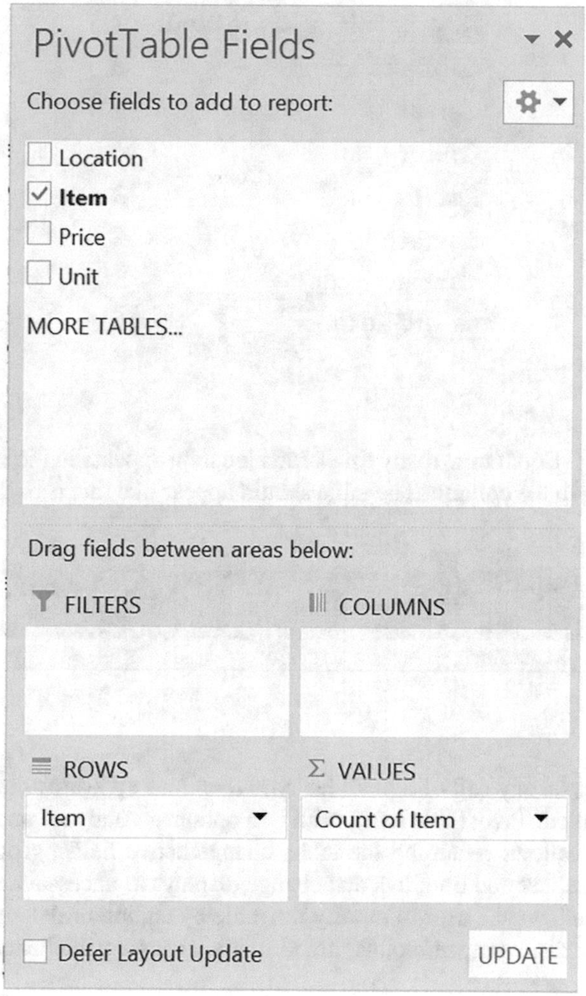

Fig. 2.146

The table will look like this.

Row Labels ▾	Count of Item
Cabbage	2
Carrot	1
Onion	4
Potato	3
Spinach	3
Tomato	3
Grand Total	**16**

Fig. 2.147

Exercise 2.16: Count how many times each location appears in the main table, the location being in the column. The table should appear like this.

	Column Labels ▾				
	East	North	South	West	Grand Total
Count of Location	3	3	7	3	16

Fig. 2.148

3. By clicking on any cell of a pivot table, two options appear available in the ribbon tab, called "PivotTable tool," with two options, "Analyze" and "Design."

    Analyze allows renaming the table, change active fields, group or ungroup data, insert slicer and timeline, and change displays as necessary.

    Design allows the user to modify the table by adding or deleting the column or rows for displaying subtotals, grand totals, changing the headers and colors with styles.

2.17 Electrical Engineering Data Visualization Through Excel Charts

It is easy to understand data through visualization. MS Excel can be used in many applications of Electrical Engineering. It is especially useful to create graphs and charts. Here are some visualizations of *electrical engineering* data drawn following the steps demonstrated throughout this chapter:

1. *Line Chart*: This is the data for the average electricity price to various types of consumers in cents per kWh.

Year	Residential	Commercial	Industrial	Transportation	All sectors
2010	11.54	10.19	6.77	10.56	9.83
2011	11.72	10.24	6.82	10.46	9.9
2012	11.88	10.09	6.67	10.21	9.84
2013	12.13	10.26	6.89	10.55	10.07
2014	12.52	10.74	7.1	10.45	10.44
2015	12.65	10.64	6.91	10.09	10.41
2016	12.55	10.43	6.76	9.63	10.27
2017	12.89	10.66	6.88	9.68	10.48
2018	12.87	10.67	6.92	9.7	10.53
2019	13.04	10.66	6.83	9.73	10.6

The line chart obtained from this data is:

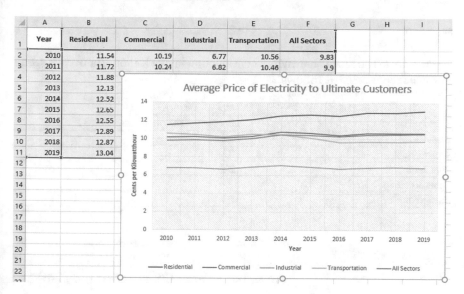

Fig. 2.149

2. *Column Chart*: The data of renewable energy consumption by the source is represented as column chart:

Source	Energy consumption (Quadrillion Btu)
Wind	2.7
Hydroelectric power	2.5
Wood	2.3
Biofuels	2.3
Solar	1
Waste	0.4
Geothermal	0.2

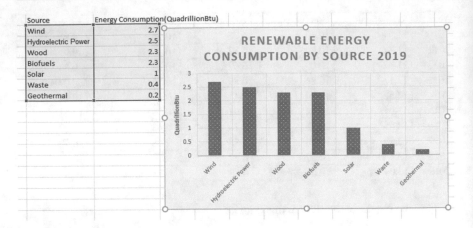

Source	Energy Consumption(QuadrillionBtu)
Wind	2.7
Hydroelectric Power	2.5
Wood	2.3
Biofuels	2.3
Solar	1
Waste	0.4
Geothermal	0.2

Fig. 2.150

3. *Pie Chart*: The renewable energy consumption in quadrillion Btu in each sector can be represented as follows:

Sector	Energy consumption (Quadrillion Btu)
Residential	0.8
Commercial	0.3
Industrial	2.5
Transportation	1.4
Electric power	6.4

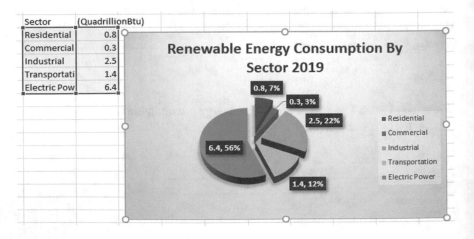

Sector	(QuadrillionBtu)
Residential	0.8
Commercial	0.3
Industrial	2.5
Transportati	1.4
Electric Pow	6.4

Fig. 2.151

4. *Bar Chart*: The yearly average price of electricity in each sector is compared with the help of a bar chart as follows:

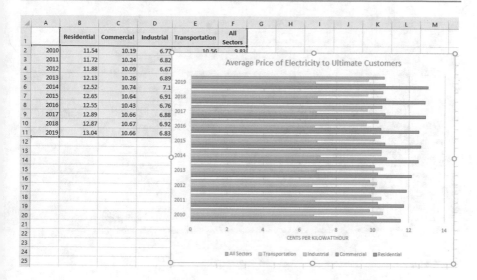

	A	B	C	D	E	F
1		Residential	Commercial	Industrial	Transportation	All Sectors
2	2010	11.54	10.19	6.77	10.56	9.83
3	2011	11.72	10.24	6.82		
4	2012	11.88	10.09	6.67		
5	2013	12.13	10.26	6.89		
6	2014	12.52	10.74	7.1		
7	2015	12.65	10.64	6.91		
8	2016	12.55	10.43	6.76		
9	2017	12.89	10.66	6.88		
10	2018	12.87	10.67	6.92		
11	2019	13.04	10.66	6.83		

Fig. 2.152

5. *Area Chart*: The renewable energy generation from 1965 to 2018 from different sources can be shown as follows:

Year	Hydropower	Wind	Solar	Others
1965	919.77	0	0	17.99
1980	1698.61	0.01	0	49.39
1990	2161.05	3.63	0.39	116.54
2000	2654.7	31.42	1.13	185.45
2010	3432.95	341.61	33.68	378.85
2018	4193.1	1269.95	584.63	625.81

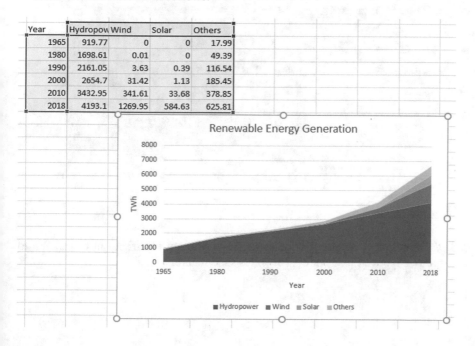

Fig. 2.153

6. *Error Bars*: The error of electric load forecasting is shown using error bars.

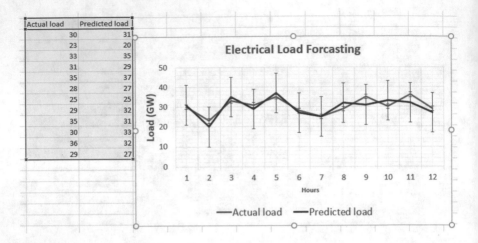

Actual load	Predicted load
30	31
23	20
33	35
31	29
35	37
28	27
25	25
29	32
35	31
30	33
36	32
29	27

Fig. 2.154

7. *Gauge Chart*: A gauge chart can be used to indicate the state of charge (SOC) of a battery.

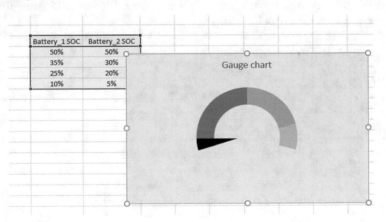

Battery_1 SOC	Battery_2 SOC
50%	50%
35%	30%
25%	20%
10%	5%

Fig. 2.155

8. *Scatter Plot*: A scatter plot of the average price of electricity in cents/kWh to various types of loads or consumers is shown here:

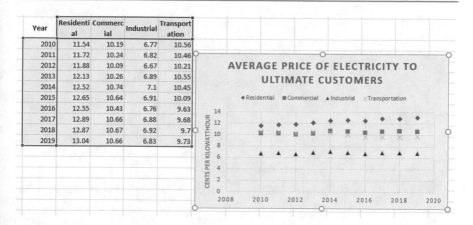

Year	Residential	Commercial	Industrial	Transportation
2010	11.54	10.19	6.77	10.56
2011	11.72	10.24	6.82	10.46
2012	11.88	10.09	6.67	10.21
2013	12.13	10.26	6.89	10.55
2014	12.52	10.74	7.1	10.45
2015	12.65	10.64	6.91	10.09
2016	12.55	10.43	6.76	9.63
2017	12.89	10.66	6.88	9.68
2018	12.87	10.67	6.92	9.7
2019	13.04	10.66	6.83	9.73

Fig. 2.156

2.18   Conclusion

Chapter 2 is undoubtedly the most important chapter of this book because it explains the most widely used features of MS Excel, i.e., Graphs and Charts. MS Excel is the go-to software everyone reaches out to for preparing graphs and charts used in academics, businesses, offices, institutions, shops, and many more. This is the feature that makes Excel so versatile and colorful and catering to the needs of all types of users. This chapter explains the process of creating different types of graphs and charts in Microsoft Excel and contains exercises to ensure proper learning of the topics. Besides, the chapter focuses on electrical engineering problems and demonstrates some graphs and charts based on electrical engineering data.

Reference

1. https://www.excel-easy.com/data-analysis/pivot-tables.html

MS Excel Functions and Formulae

3

Learning Objectives
- Learn about the formats of basic functions.
- Use functions to build formulae.
- Perform basic arithmetic and logical operation using Microsoft Excel.

3.1 Excel Functions and Formulas

So far, it is evident that Microsoft Excel is an excellent tool for data manipulation, which could be very tiresome if done by hand. Excel functions are the keywords with the purpose to facilitate various operations according to the topic or subjects. For example, if one were to add to numbers stored in cells A1 and A2, one would need to use a function called SUM in a particular format to sum the data in those cells. Several complex operations of ranging categories are possible with the help of functions. The format or expression of the statement to execute an operation is called the formula. That means a formula should be accompanied by a function because the function will make the formula work with its embedded code in Excel. The terms are sometimes interchangeably used in the context, as the formula must include a function. In this book, the keywords or functions will be introduced with examples of formulas for the clarification of the readers.

Microsoft Excel functions are accessible using the following options.

1. Click on the "Formulas" option of the ribbon tab.

© The Author(s), under exclusive license to Springer Nature Switzerland AG 2021
E. Hossain, *Excel Crash Course for Engineers*,
https://doi.org/10.1007/978-3-030-71036-1_3

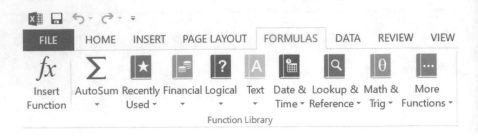

Fig. 3.1

On clicking the "Formulas" option, a number of function libraries appear. By selecting the required category, the formula can be utilized in the selected cell.

2. By manually typing the function in the formula bar/cell

The formula is followed by an equal sign, then by the function. The values of the functions are bounded by the first bracket (). In order to add two numbers, 2 and 3, in cell A1, the formula can be constructed as

$$= SUM(2,3)$$

AND	▾	⋮	×	✓	f_x	=SUM(2,3)

	A	B	C	D	E
1	=SUM(2,3)				
2					

Fig. 3.2

On pressing "Enter," the value 5 will appear.

3.2   Basic Excel Functions

The following table enlists some basic functions available in MS Excel:

Basic Excel functions	More functions
Financial	Statistical
Logical	Engineering
Text	Cube
Date & time	Information
Lookup & reference	Compatibility
Math & trigonometry	Web

Excel tries to make it easier by creating a Recently Used menu that contains a list of the Functions one has used most often. From this option, one will be able to view a list of the functions that one uses the most and will easily be able to use one of the functions without searching from the different function library menus.

3.2.1   Logical Functions

Excel has five major logical functions to operate on binary or logical values. They are NOT, AND, IF, OR, & XOR.

NOT

Summary: Outputs the inverse of a given Boolean or logical value. It returns FALSE when a TRUE value is given and vice versa.
Syntax: = NOT(logical)
Argument: Logical – Any input that can be categorized as TRUE or FALSE.
Example:

A1		▾	⋮	✕	✓	*fx*	=NOT(1)

	A	B	C	D
1	FALSE			
2				

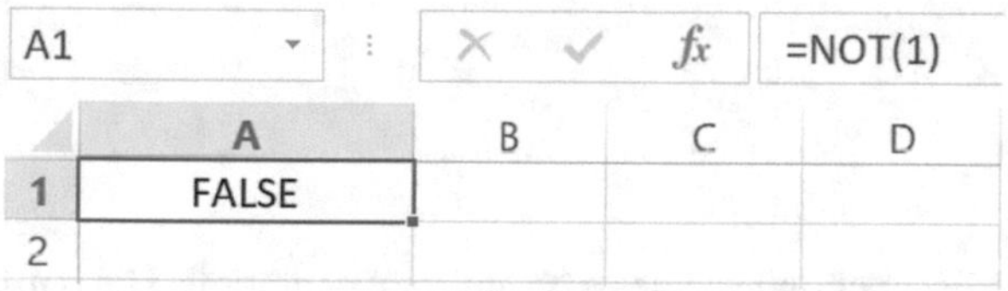

Fig. 3.3

On typing =NOT(1) will return FALSE, as the opposite of 1 is 0. Similarly, =NOT(0) or =NOT(FALSE) will return TRUE.

AND

Summary: AND function checks the arguments and outputs TRUE if all the arguments are TRUE or 1, and yields FALSE if any of the arguments is FALSE or 0.
Syntax: =AND(logical1, [logical2], …)
Argument: Logical – Any input that can be categorized as TRUE or FALSE.
Example:

B4		▾	⋮	✕	✓	*fx*	=AND(A4>0,A4<100)

	A	B	C	D	E
1					
2	Return TRUE if value is >0 and <100				
3					
4	50	TRUE			
5	70				
6	100				
7	150				
8	23				
9					

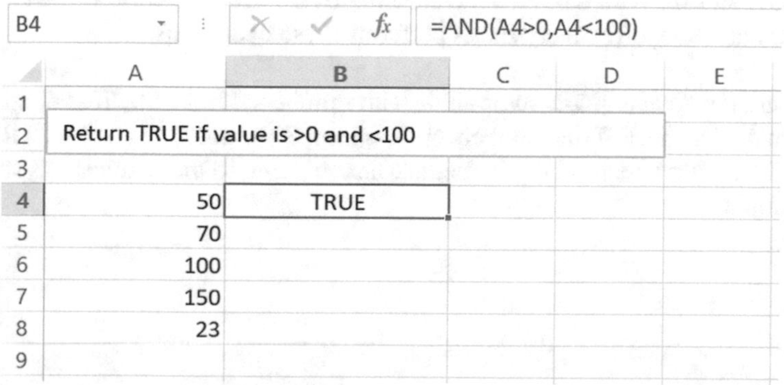

Fig. 3.4

For the rest of the values, the result can be found as follows:

B4	▾	⋮	✕	✓	*fx*	=AND(A4>0,A4<100)	

◢	A	B	C	D
1				
2	Return TRUE if value is >0 and <100			
3				
4	50	TRUE		
5	70	TRUE		
6	100	FALSE		
7	150	FALSE		
8	23	TRUE		
9				
10				

Fig. 3.5

IF

Summary: The IF function executes a conditional operation. The output of the function depends on the input. If the input is TRUE, then the output is different from the output if the input is FALSE. It means the IF statement can show two results.
Syntax: =IF (condition1, [value_if_true], [value_if_false])
Example: The IF function can be used alongside the logical functions such as AND and OR.

OR

Summary: OR function checks the arguments and outputs TRUE if any of the arguments are TRUE or 1, and FALSE if all of the arguments is FALSE or 0.
Syntax: =OR (logical1, [logical2], …)

Insert 0s and 1s randomly in Excel, and play around with the functions AND, OR, NOT, and IF. With a basic understanding of Boolean Algebra, you will be able to understand how effectively Excel performs these operations.

Exercise 3.1: Create a 5-column-table with numbers 23, 36, 90, 78, 56, and using function AND, check if the numbers are less than 50 or not. The answer "TRUE" or "FALSE" should appear in the immediate next column to the numbers, as shown in the example.

3.2.2 Date & Time Functions

In order to write a specific date, use the characters "/" or "-." For entering the time, the ":" (colon) can be used. A date and a time can be inserted into a single cell [1].

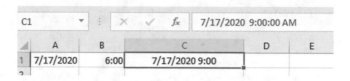

Fig. 3.6

Note: Dates are in the U.S. Format, i.e., Month/Date/Year. This format depends on the regional settings on Windows.

Year, Month, and Day
Use the YEAR function for knowing the year from a date.

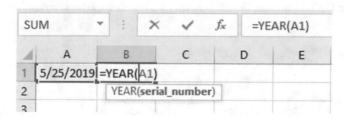

Fig. 3.7

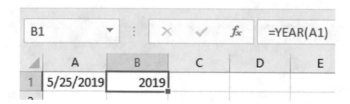

Fig. 3.8

Note: To know the month and day from a date, use the MONTH and DAY function.

Date Function

1. Excel allows us to add a number of days to a certain date. Write the cell name containing the desired date and add to it the number of days you want.

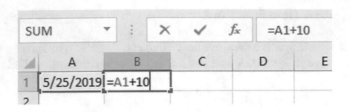

Fig. 3.9

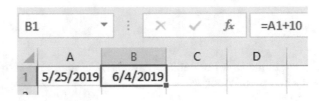

Fig. 3.10

2. Use the DATE function to get the sum of a number of days, months, or years.

SUM	▼	:	✕	✓	*fx*	=DATE(YEAR(A1)+2,MONTH(A1)+3,DAY(A1)+10)			
	A		B			C	D	E	F
1	5/25/2019	=DATE(YEAR(A1)+2,MONTH(A1)+3,DAY(A1)+10)							
2									

Fig. 3.11

B1	▼	:	✕	✓	*fx*	=DATE(YEAR(A1)+2,MONTH(A1)+3,DAY(A1)+10)		
	A	B	C	D	E	F	G	H
1	5/25/2019	9/4/2021						
2								

Fig. 3.12

The DATE function has three arguments: year, month, and day. In Excel, the information about the number of days of each month and year, and how the calendar works are in-built. So, it is needless to doubt the accuracy of these calculations in Excel.

Current Date & Time
The NOW function can be used to write the current date and time, based on the user's location.

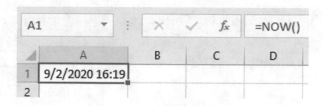

Fig. 3.13

Note: You can use the TODAY function for entering the present date. The TODAY function has no argument.

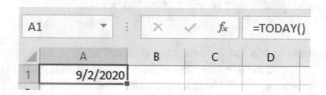

Fig. 3.14

HOUR, MINUTE, SECOND

Use the HOUR function for inserting the hour of a certain day. The hour is displayed in the 24-hour clock format, i.e., from 0 to 23. The HOUR function takes in a decimal value as the input, referring to what fraction of the day it is, and outputs the hour as an integer value. For example, =HOUR(0.5) will return 12, since 12 o'clock is exactly one-half of 24 hours. Similarly, =HOUR(0.75) will return 18, which is 3-quarters of the 24-hour span.

B1		⋮	× ✓ f_x	=HOUR(A1)		
	A		B	C	D	
1	9/2/2020 16:23		16			

Fig. 3.15

Exercise 3.2: Use the MINUTE and SECOND functions to write the minute and second in Excel.

TIME Function

Use the TIME function to display the sum of a number of hours, minutes, and/or seconds.

B1	▾	:	×	✓	*fx*	=TIME(HOUR(A1)+9,MINUTE(A1)+20,SECOND(A1)+20)

◢	A	B	C	D	E	F	G
1	16:23:45	1:44 AM					

Fig. 3.16

3.3   Math and Trigonometry

MS Excel has more than 70 functions to work on mathematical and trigonometric operations. Some notable functions are described below:

ABS
The ABS function in MS Excel outputs the absolute value of a number. For example, in Fig. 3.17, cell B1 contains the absolute value of cell A1 by the formula ABS(A1). Similarly, in Fig. 3.18, cell B2 contains absolute value of cell A2 by the formula ABS(A2). The absolute value of both the numbers 12 and -12 is 12.

B1	▾	:	×	✓	*fx*	=ABS(A1)

◢	A	B	C	D
1	12	12		
2	-12	12		

Fig. 3.17

B2	▾	:	×	✓	*fx*	=ABS(A2)

◢	A	B	C	D
1	12	12		
2	-12	12		
3				

Fig. 3.18

BASE
The BASE function in Excel converts a decimal number into a number in another number system, defined by the given base or radix. For example, in Fig. 3.19, the BASE function converts the decimal number 255 into its binary equivalent

11111111. The number is given in cell B1, the radix is also given in cell B2, and the output is produced in cell B3 by using the formula =BASE(B1,B2). The syntax is:

=BASE(number,radix,min_length).

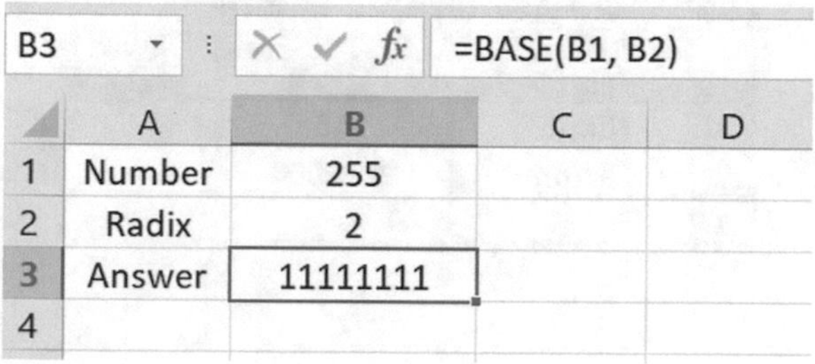

Fig. 3.19

The desired minimum length of the output number can be obtained by padding the result with zeros. For instance, in Fig. 3.20, the formula is inserted as =BASE(B1,B2,10). The output is a 10-digit number padded with zeros.

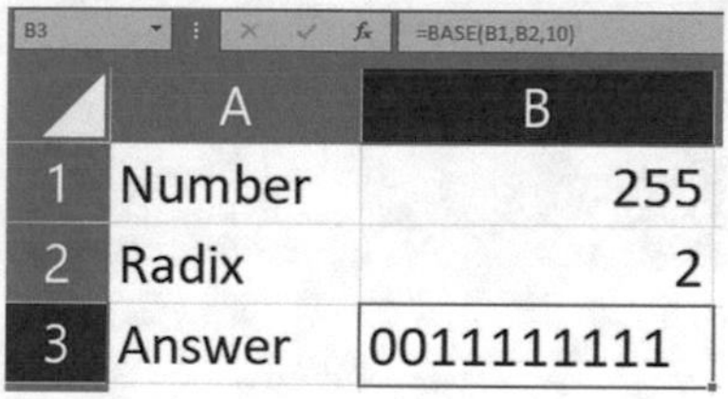

Fig. 3.20

EXP

The EXP function outputs the exponential value of a number, i.e., the power of the exponential constant $e = 2.718$. For example, Fig. 3.21 demonstrates the exponential value of 2, i.e., the value of the constant e raised to the power of 2, in the cell B2 by the use of the formula =EXP(B1).

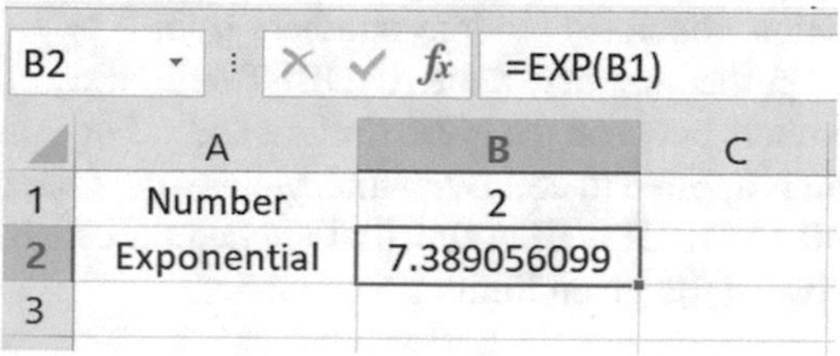

Fig. 3.21

FACT

The FACT function is used to get the factorial of a given number. For example, in Fig. 3.22, the cell B2 contains the factorial of the number in cell B1, i.e., 10! upon using the formula =FACT(B1).

B2	▾	:	✕ ✓ *fx*	=FACT(B1)

◢	A	B	C
1	Number	10	
2	Factorial	3628800	
3			

Fig. 3.22

Trigonometric Functions

As trigonometric functions shown below takes values in radian, the desired degree values are converted to radian first with RADIANS function, then each of the functions mentioned in the first columns is calculated to determine their normal and hyperbolic value, as visible from the following table.

	A	B	C	D	E	F	G	H	I	J	K	L	M	N
1	ANGLES IN DEGREES	ANGLES IN RADIANS	SIN()	SINH()	COS()	COSH()	TAN()	TANH()	CSC()	CSCH()	SEC()	SECH()	COT()	COTH()
2	0	0	0	0	1	1	0	0	#DIV/0!	#DIV/0!	1	1	#DIV/0!	#DIV/0!
3	30	0.523598776	0.5	0.547853	0.866025	1.140238	0.57735	0.480473	2	1.825306	1.154701	0.87701	1.732051	2.081283
4	45	0.785398163	0.707107	0.868671	0.707107	1.324609	1	0.655794	1.414214	1.151184	1.414214	0.75494	1	1.524869
5	60	1.047197551	0.866025	1.249367	0.5	1.600287	1.732051	0.780714	1.154701	0.800405	2	0.624888	0.57735	1.280878
6	90	1.570796327	1	2.301299	6.13E-17	2.509178	1.63E+16	0.917152	1	0.434537	1.63E+16	0.398537	6.13E-17	1.090331

Fig. 3.23

Exercise 3.3: Why do you think the "#DIV/0!" error occurred for cosecant, hyperbolic cosecant, cot, and hyperbolic cotangent?

The inverse trigonometric functions such as ASIN(), ASINH(), ACOS(), ACOSH(), ATAN(), ATANH(), ACOT(), and ACOTH() all follow the same syntax where the argument is a number and the functions return an angle in radians.

Random Numbers

The RAND() function generates random numbers from 0 to 1. These are always decimal numbers. On the contrary, the RANDBETWEEN(x,y) function generates random integer numbers between the given limits x and y. For example, in Fig. 3.24, the RAND function is applied to cell B2, which yields a decimal number between 0 and 1. Again, the RANDBETWEEN function applied on cell B2 produces 2 as a random integer between the given limits 1.5 and 2.5.

B2	▾	⋮	✕	✓	*fx*	=RANDBETWEEN(1.5,2.5)	

◢	A	B	C
1	RAND Function	0.966272265	
2	RANDBETWEEN Function	2	
3			

Fig. 3.24

Some other mentionable Excel functions are briefly discussed below.

SQRT. The SQRT function in Excel produces the square root of a given number. However, the SQRT function returns an error if the input number is negative. The syntax is:

=SQRT(number).

GCD. The GCD function in Excel produces the Greatest Common Divisor (GCD) of the given numbers. The number of arguments in this function may be two or more. The syntax is:

=GCD(number1, number2, number3, ...)

LCM. The LCM function in Excel produces the Least Common Multiple (LCM) of the given numbers. The number of arguments in this function may be two or more. The syntax is:

=LCM(number1, number2, number3, ...)

LOG. The LOG function in Excel produces the logarithmic value of a given number to the base specified in the arguments. The syntax is:

=LOG(number, base)

LOG10. The LOG10 function in Excel produces the logarithmic value of a given number to the base 10. The syntax is:

=LOG(number)

More Functions.   These are more technical functions for specific purposes such as statistics, engineering, cubic operation, information analysis, compatibility testing, and web interfacing. Although the application of statistics, compatibility functions are mentioned in Sect. 5.8.

3.4   Engineering Functions

MS Excel consists of more than 50 functions, which are more focused on mathematical operations for engineering applications. Some of these functions are categorized below as per their usage.

Bessel's Function
Bessel's function calculation can be done in MS Excel for engineering applications, especially for analyzing wave characteristics. BESSELJ and BESSELY return the Bessel's function, and BESSELI and BESSELK return the modified Bessel's function. All the values are determined for $x = 1$ and $n = 5$, where x is the point at which the Bessel's function is to be determined, and n is the order of the Bessel's function, which must be a positive integer.

BESSELJ denotes the Bessel's function $J_n(x)$, while the BESSELY denotes the Weber function (also called Neumann function) $Y_n(x)$. BESSELI and BESSELK denote the modified Bessel's function (also called hyperbolic Bessel's function) $I_n(x)$ and $K_n(x)$, respectively. For each of these functions, the syntax is the same as:

=BESSELY(x, n), or BESSELI(x, n), or BESSELJ(x, n), or BESSELK(x, n)

B4	▾	⋮	✕	✓	*fx*	=BESSELY(1,5)

	A	B
1	BESSELI function	0.000271463
2	BESSELJ function	0.000249758
3	BESSELK function	360.9605868
4	BESSELY function	-260.4058678

Fig. 3.25

Base Conversion
The base of a number, i.e., binary (BIN), decimal (DEC), octal (OCT), and hexadecimal (HEX), can be changed using Excel functions such as BIN2DEC, BIN2HEX, BIN2OCT, HEX2BIN, HEX2OCT, HEX2DEC, OCT2BIN, OCT2DEC,

and OCT2HEX. Figure 3.26 demonstrates the conversion from decimal to binary, octal, hexadecimal using DEC2BIN, DEC2OCT, and DEC2HEX functions.

	A	B	C	D
1	DEC	BIN	OCT	HEX
2	0	0	0	0
3	1	1	1	1
4	2	10	2	2
5	3	11	3	3
6	4	100	4	4
7	5	101	5	5
8	6	110	6	6
9	7	111	7	7
10	8	1000	10	8
11	9	1001	11	9
12	10	1010	12	A
13	11	1011	13	B
14	12	1100	14	C
15	13	1101	15	D
16	14	1110	16	E
17	15	1111	17	F
18	16	10000	20	10

Fig. 3.26

Bit Operations

MS Excel offers a number of binary operations through the functions BITLSHIFT, BITRSHIFT, BITAND, BITOR, BITXOR, etc. The BITLSHIFT and BITRSHIFT functions are similar and have the same syntax:

=BITLSHIFT(number, shift_amount), or BITRSHIFT(number, shift_amount).

	A	B	C	D	E	F	G
1	Number	Binary Representation	Shift_Amount	After Right Shift	BITRSHIFT function	After Left Shift	BITLSHIFT function
2	5	101	0	101	5	101	5
3	6	110	1	11	3	1100	12
4	7	111	2	1	1	11100	28
5	8	1000	3	1	1	1000000	64
6	9	1001	4	0	0	10010000	144

Fig. 3.27

These functions shift the binary number to the left or right by adding 0 on the emptied place of the previous bits on the leftmost or rightmost side of the number.

The number of digits removed is specified by the shift_amount. For instance, consider the decimal number 10, whose binary representation is 1010. Say we are to shift this number by 2 places to the left. The result is 101000, which is equivalent to the decimal number 40. Notice that the two empty places are padded with two 0s. Again, if we shift the number 10 by 2 places to the right, we obtain 0010, which is equal to decimal 2. The rightmost "10" disappears due to the absence of a place value. The leftmost two empty places are filled in with 0s. So, BITLSHIFT(10,2)=40 and BITRSHIFT(10,2)=2. Several other examples are demonstrated in Fig. 3.27, with a detailed breakdown of the numbers.

◢	A	B
1	A	10101
2	B	10010
3	BITAND(A,B) function	10000
4	BITOR(A,B) function	10111
5	BITXOR(A,B) function	111

Fig. 3.28

The BITAND(number1, number2, ...) function performs the AND operation on two binary numbers, the BITOR(number1, number2, ...) function performs the OR operation on two binary numbers, and the BITXOR(number1, number2, ...) function performs the XOR operation on two binary numbers. The implementation of all these functions upon two binary numbers A=10101 and B=10010 is demonstrated in Fig. 3.28.

Imaginary Numbers (Basic Mathematical Operation)
MS Excel contains functions for handling imaginary numbers too. Some of these functions are detailed below and implemented in Fig. 3.29.

The *COMPLEX function* is used to denote a number in the complex notation $a \pm bi$. The syntax is:

=COMPLEX(real_part, imaginary_part, "suffix")

Excel allows the usage of both i and j as the imaginary operator, which can be specified by the suffix argument inside double quotation marks. The function works even without the suffix argument, and the default operator, in that case, is i.

The *IMABS*(number) function outputs the absolute value of a complex number. For the complex number $a \pm bi$, this function returns the value $\sqrt{a^2 + b^2}$.

The *IMREAL*(number) function outputs the real part of a complex number. For the complex number $a \pm bi$, this function returns the value a.

The *IMAGINARY*(number) function outputs the imaginary part of a complex number. For the complex number $a \pm bi$, this function returns the value $\pm b$.

The *IMARGUMENT*(number) function outputs the argument, or angle, made by a complex number in the polar coordinate system.

The *IMCONJUGATE*(number) function outputs the complex conjugate form of the given number. For the complex number $a + bi$, this function returns the value $a - bi$.

The *IMSUM*(number1, number2, ...) function outputs the sum of two or more complex numbers.

The *IMSUB*(number1, number2) function outputs the difference of two complex numbers by performing the operation number1 − number2.

The *IMPRODUCT*(number1, number2, ...) function outputs the product of 1–255 complex numbers.

The *IMDIV*(number1, number2) function outputs the quotient of the division of complex number 1 by complex number2.

The *IMSQRT*(number) function outputs the square root of the given complex number. For the complex number $a \pm bi$, this function returns the value $\sqrt{(a \pm bi)}$.

The *IMPOWER*(cx_number, number) outputs a complex number raised to the given integer power.

The *IMEXP*(number) function outputs the exponential of a complex number. For the complex number $a \pm bi$, this function returns the value $e^{(a \pm bi)}$.

The *IMLOG2*(number) function outputs the logarithm of a complex number to the base 2. For the complex number $a \pm bi$, this function returns the value $\log_2(a \pm bi)$.

The *IMLOG10*(number) function outputs the logarithmic value of a complex number to the base 10. For the complex number $a \pm bi$, this function returns the value $\log_{10}(a \pm bi)$.

	A	B	C
1	A = 1+2i	COMPLEX(1,2)	1+2i
2	B = 4-5i	COMPLEX(4,-5)	4-5i
3	Modulus of A	IMABS(C1)	2.236067977
4	Argument of B	IMARGUMENT(C2)	-0.896055385
5	Complex Coefficient of A	IMAGINARY(C1)	2
6	Conjugate of B	IMCONJUGATE(C2)	4+5i
7	Sum of A and B	IMSUM(C1,C2)	5-3i
8	Subtraction of B from A	IMSUB(C1,C2)	-3+7i
9	Multiplication of A and B	IMPRODUCT(C1,C2)	14+3i
10	Division of B from A	IMDIV(C1,C2)	-0.146341463414634+0.317073170731707i
11	Squareroot of A	IMSQRT(C1)	1.27201964951407+0.786151377757423i
12	Square of A	IMPOWER(C1, 2)	-3+4i
13	Exponential of B	IMEXP(C2)	-1.13120438375681+2.47172667200482i
14	Log2 of B	IMLOG2(C2)	2.67877600230904-1.29273465968293i

Fig. 3.29

Imaginary Numbers (Trigonometric Operation)

MS Excel has several functions for trigonometric operations on complex numbers. All these functions take in one complex number as input and return the trigonometric function of that number. Examples of such functions are IMSIN, IMCOS,

IMTAN, IMSINH, IMCOSH, IMSEC, IMSECH, IMCSC, IMCSCH, IMCOT, etc., and these are implemented in Fig. 3.30.

	A	B	C
1	Z = 5-6i	COMPLEX(5,-6)	5-6i
2	Sine of Z	IMSIN(C1)	-193.43002005694-57.2183950563411i
3	Cosine of Z	IMCOS(C1)	57.2190981846007-193.427643121306i
4	Tangent of Z	IMTAN(C1)	-6.68523139027702E-06-1.00001031089812i
5	Hyperbolic Sine of Z	IMSINH(C1)	71.2477179708529+20.7354097383702i
6	Hyperbolic Cosine of Z	IMCOSH(C1)	71.2541875473544+20.7335270515526i
7	Secant of Z	IMSEC(C1)	0.00140627965955837+0.00475389107392496i
8	Hyperbolic Secant of Z	IMSECH(C1)	0.0129387490225521-0.00376491420513424i
9	Cosecant of Z	IMCSC(C1)	-0.00475385145844243+0.00140623337943227i
10	Hyperbolic Cosecant of Z	IMCSCH(C1)	0.0129395569922043-0.00376583312010237i
11	Cotangent of Z	IMCOT(C1)	-6.68509353063089E-06+0.999989689163503i

Fig. 3.30

3.5   Application of Excel Functions in Electrical Engineering

3.5.1   Ohm's Law

Voltage (V), current (I), and resistance (R) are the three fundamental parameters of electricity. Ohm's law establishes a relationship among these three quantities. The law states that at a particular temperature, the current flowing through a conductor connecting two points is directly proportional to the difference of voltage across those two points. This relationship is depicted in Fig. 3.31.

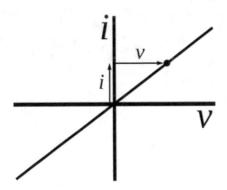

Fig. 3.31

According to Ohm's law, $V = I \times R$. The units of voltage, current, and resistance are respectively Volt (V), Ampere (A), and Ohm (Ω).

Applications of Ohm's Law

Ohm's law is a rudimentary law of electricity, facilitating the calculation of several parameters in electric calculations. We can use the equation of Ohm's law for calculating either voltage, current, or resistance if the other two quantities are known. It also makes other complex calculations much easier. It aids in power calculations, too [2].

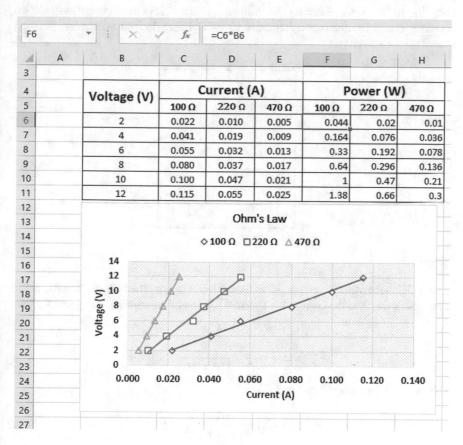

F6		×	✓	f_x	=C6*B6		

	A	B	C	D	E	F	G	H
3								
4		**Voltage (V)**	**Current (A)**			**Power (W)**		
5			100 Ω	220 Ω	470 Ω	100 Ω	220 Ω	470 Ω
6		2	0.022	0.010	0.005	0.044	0.02	0.01
7		4	0.041	0.019	0.009	0.164	0.076	0.036
8		6	0.055	0.032	0.013	0.33	0.192	0.078
9		8	0.080	0.037	0.017	0.64	0.296	0.136
10		10	0.100	0.047	0.021	1	0.47	0.21
11		12	0.115	0.055	0.025	1.38	0.66	0.3

Fig. 3.32

The formulae and functions in MS Excel can be used for analyzing data obtained from electric and electronic circuits to interpret Kirchoff's voltage and current law, from Thevenin and Norton's equivalent circuits to be used in numerous applications such as maximum power transfer theorem. Since Excel does not perform simulation, we have to solve circuits by other software, or manually, the data from which can be represented in Excel for easy display and modification by the academicians or for performing trial and error by students or researchers.

3.5.2   Maximum Power Transfer Theorem

From the theory, it is evident that no matter how the circuit is, to determine the maximum power that passes through the load, we need to determine the Thevenin voltage, V_{TH}, and the Thevenin resistance R_{TH} after constructing the Thevenin equivalent circuit. The circuit is composed of a voltage source rated V_{TH} and a resistor rated R_{TH} in series. Considering the following basic series circuit, consisting of two resistors Ri and RL and one voltage source VCC, being derived from a complicated circuit after analysis:

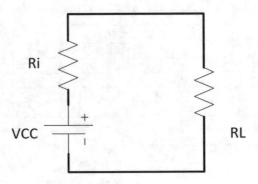

Fig. 3.33

If VCC is the voltage source, Ri is the circuit's internal resistance, and RL is the load resistance, then the current I in the load RL can be determined using Ohm's law:

$$V = IR$$

$$\Rightarrow I = \frac{V}{R} = \frac{V}{(Ri + RL)}.$$

Now, to determine the power P,

$$P = (I)^2 (R) = \frac{V^2}{(Ri + RL)^2} * (RL).$$

Considering VCC = 12 V and Ri = 2 Ohm, the value of power P can be determined from a range of load resistance RL using Microsoft Excel. Let us tabulate the load resistance ranging from 1 to 20 Ohm.

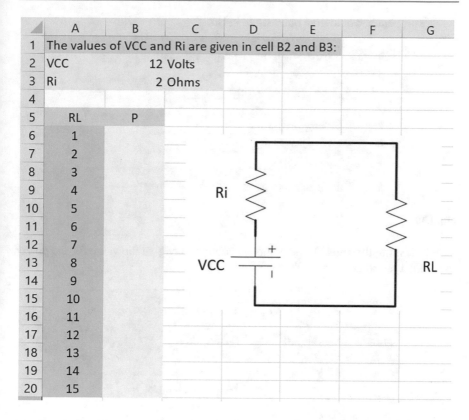

Fig. 3.34

Here, the column P (from cell B6 to B20) is required to be populated with the value of power from the formula. Let us put the following formula in cell B6:

$$P = \frac{V^2}{(Ri+RL)^2} * RL = \frac{(\text{Data in cell B2})^2}{\{(\text{Data in cell B3}) + (\text{Data in cell A6})\}^2} * (\text{Data in cell A6}).$$

It will give out the value of the power in cell B6 for RL = 1 Ohm only.

COR... ▾	⋮	✗ ✓ f_x	=((B2*B2)/((B3+A6)*(B3+A6)))*A6

	A	B	C	D	E	F
1	The values of VCC and Ri are given in cell B2 and B3:					
2	VCC	12	Volts			
3	Ri	2	Ohms			
4						
5	RL	P				
6	1	=((B2*B2)/((B3+A6)*(B3+A6)))*A6				
7	2					

Fig. 3.35

Pressing enter would give out a value in B6.

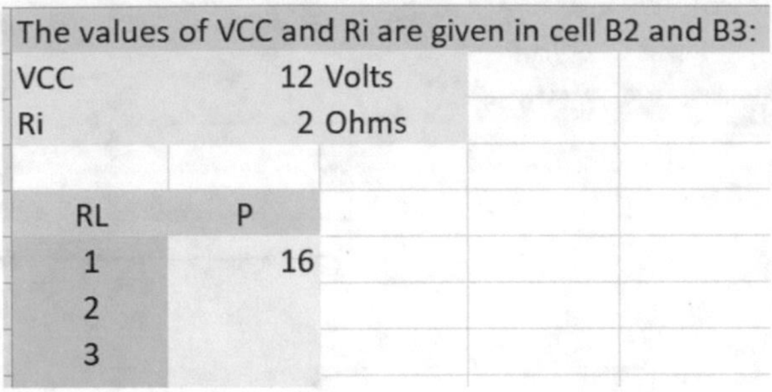

The values of VCC and Ri are given in cell B2 and B3:		
VCC	12 Volts	
Ri	2 Ohms	
RL	P	
1	16	
2		
3		

Fig. 3.36

Fill-flashing the value in the complete column may give the value for the power in the load resistance.

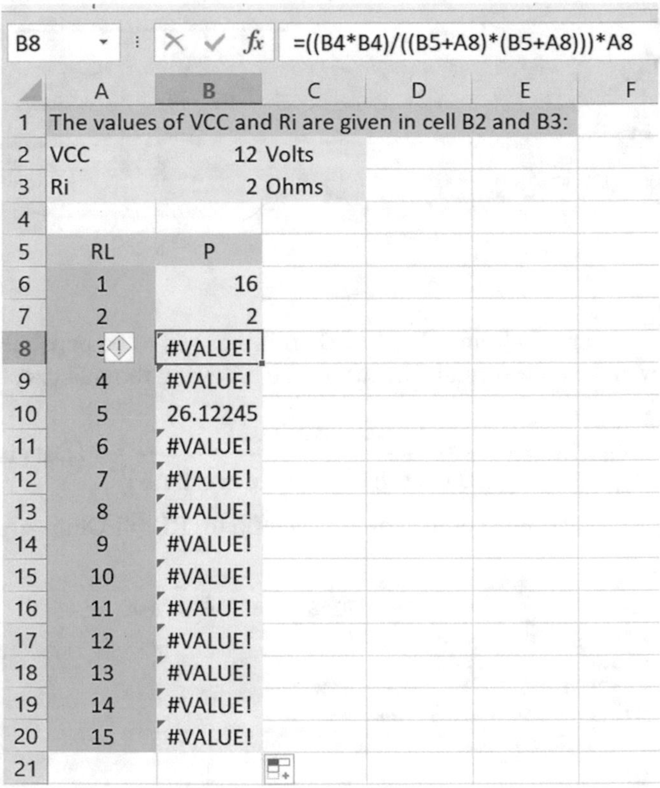

B8		× ✓ f_x	=((B4*B4)/((B5+A8)*(B5+A8)))*A8			
	A	B	C	D	E	F
1	The values of VCC and Ri are given in cell B2 and B3:					
2	VCC	12 Volts				
3	Ri	2 Ohms				
4						
5	RL	P				
6	1	16				
7	2	2				
8	3	#VALUE!				
9	4	#VALUE!				
10	5	26.12245				
11	6	#VALUE!				
12	7	#VALUE!				
13	8	#VALUE!				
14	9	#VALUE!				
15	10	#VALUE!				
16	11	#VALUE!				
17	12	#VALUE!				
18	13	#VALUE!				
19	14	#VALUE!				
20	15	#VALUE!				
21						

Fig. 3.37

But this gives an error from Cell B8. From the formula above, it is noticeable that not only the values of RL get modified due to fill-flash, but also the values of B2 and

B3 get iterated due to the innate functionality of fill-flash. Therefore, each subsequent cell shows an error.

To remove the errors, the cells with the values of VCC and Ri are considered as absolute cells. The formula is modified in cell B6 as follows.

COR...	⯆	⋮	✕ ✓ ƒx	=((B$2*B$2)/((B$3+A6)*(B$3+A6)))*A6		

◢	A	B	C	D	E	F
1	The values of VCC and Ri are given in cell B2 and B3:					
2	VCC	12	Volts			
3	Ri	2	Ohms			
4						
5	RL	P				
6	1	=((B$2*B$2)/((B$3+A6)*(B$3+A6)))*A6				
7	2	2				
8	3	#VALUE!				
9	4	#VALUE!				

Fig. 3.38

Pressing enter would show the same value, but this time, fill-flash would work correctly as cell B2 and B3 will remain fixed no matter what.

B6	⯆	⋮	✕ ✓ ƒx	=((B$2*B$2)/((B$3+A6)*(B$3+A6)))*A6		

◢	A	B	C	D	E	F
1	The values of VCC and Ri are given in cell B2 and B3:					
2	VCC	12 Volts				
3	Ri	2 Ohms				
4						
5	RL	P				
6	1	16				
7	2	18				
8	3	17.28				
9	4	16				
10	5	14.69388				
11	6	13.5				
12	7	12.44444				
13	8	11.52				
14	9	10.71074				
15	10	10				
16	11	9.372781				
17	12	8.816327				
18	13	8.32				
19	14	7.875				
20	15	7.474048				
21						

Fig. 3.39

If RL = Ri = 2, maximum power of 18 W is obtained from the circuit. This can also be verified if the Ri is changed to 7 Ohm, and the highest power is found to be 5.14 W.

	A	B	C	D	E
1	The values of VCC and Ri are given in cell B2 and B3:				
2	VCC	12 Volts			
3	Ri	7 Ohms			
4					
5	RL	P			
6	1	2.25			
7	2	3.555556			
8	3	4.32			
9	4	4.760331			
10	5	5			
11	6	5.112426			
12	7	5.142857			
13	8	5.12			
14	9	5.0625			
15	10	4.982699			
16	11	4.888889			
17	12	4.786704			
18	13	4.68			
19	14	4.571429			
20	15	4.46281			

Fig. 3.40

The data can also be plotted in a scatter chart in smooth lines and markers. The diagram would appear like this.

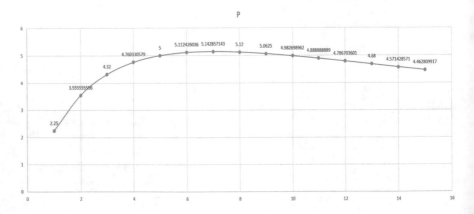

Fig. 3.41

Exercise 3.4: If a circuit has a source voltage and internal resistance of

(a) 24 V and 6 ohm,
(b) 40 V and 6 ohm,
(c) 5 V and 0.5 ohm,

respectively, then determine the resistance at which maximum power would occur? What is the value of the maximum power? With a scatter plot, show the power trend with respect to resistance.

3.5.3   Power Factor Correction

In electric systems, the power factor (pf) should be high, the ideal value being 1. The pf ranges from 0 to 1. The pf is lowered due to inductive loads and can be improved by connecting capacitor banks. Power factor connection is a method to improve the power factor by lessening the amount of reactive power of AC flow. This enhances the system efficiency and reduces the current [3, 4]. There are two components of an AC power flow, viz. Real or Active power (P) and Reactive power (Q). Together, these two components comprise the Complex power (S), whose magnitude is known as Apparent power.

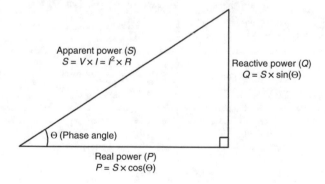

Fig. 3.42

The following formulae are adopted to calculate the power factor. Here, P = real power, Q = reactive power, and S = apparent power. The power factor is usually denoted by pf in all formal and practical cases.

$$S = P + jQ.$$

$$|S|^2 = P^2 + Q^2.$$

$$|S| = \sqrt{P^2 + Q^2}.$$

$$\text{Power factor}, \text{pf} = \cos\theta = \frac{\text{Real power}, P}{\text{Apparent power}, |S|} = \cos\left(\arctan\left(\frac{Q}{P}\right)\right).$$

$$\text{Reactive power}, \; Q = P \times \tan\left(\arccos(\text{pf})\right).$$

An Example of Power Factor Correction

A single-phase motor is connected to a supply rated 400 V, 50 Hz. The motor absorbs a current of 31.7 A at a pf of 0.7 lagging. What is the capacitance required parallel to the motor, so as to raise the pf to 0.9 lagging?

J16	▼	:	×	✓	f_x		

	A	B	C	D	E	F
1						
2						
3		Motor Current(A)=	I_M =	31.7		
4		Voltage(V)=	V=	400		
5		Initial Power Factor=	$\cos\theta_M$=	0.7	$\sin\theta_M$=	0.714143
6		Frequency(Hz)=	f=	50		
7		Corrected Power Factor=	$\cos\theta$=	0.9	$\sin\theta$=	0.43589
8		Aparent Power(kVA), S=	12.68			
9						
10		Active Component of I_{MA}=	$I_M*\cos\theta_M$=	22.19		
11		Current(A) I=	$I_M/\cos\theta$	24.65555556		
12		Active Component of I_A=	$I*\cos\theta$	22.19		
13		True Power(kW) at 0.7 p.f, P=	8.876			
14						
15						
16		Reactive Component of I_{MR}=	$I_M*\sin\theta_M$=	22.63832812		
17		Reactive Component of I_R=	$I*\sin\theta$=	10.74710751		
18						
19						
20		Capacitor Current(A), I_c=	V/X_c=	$V*2\pi*f*C$	I_{MR}-I_R	11.89122
21		Capacitance(F), C=	$I_c/(V*2\pi*f)$=	9.46753E-05		
22		Capacitance(μF), C=	94.68			
23		True Power(kW) at 0.9 p.f, P=	11.412			
24						
25		Power Factor with the Capacitor=	P/S =	0.9		

Fig. 3.43

The calculation reveals the value of the capacitor to be 94.68 micro Farads.

3.6   Conclusion

This chapter addresses the functions and formulae used in Microsoft Excel. Functions and formulae make tedious works very easy and fast in Excel. Some basic Excel functions, such as logical AND, OR, NOT, IF, and XOR, and various time and date functions are described here with their respective syntax and examples. A thorough study of this chapter will enable readers to use these functions with ease. This chapter also demonstrates the use of some Excel functions in electrical engineering. Examples of Ohm's law and power factor correction are provided for a clear illustration. This chapter will be a valuable reference for all Excel users, and especially for those who seek expertise in this program.

References

1. https://www.excel-easy.com/functions/date-time-functions.html
2. https://www.toppr.com/guides/physics/electricity/ohms-law-and-resistance/
3. https://www.allaboutcircuits.com/textbook/alternating-current/chpt-11/practical-power-factor-correction/
4. https://en.wikipedia.org/wiki/Power_factor

MS Excel VBA

4

Learning Objectives
- Learn about Visual Basic for Microsoft Excel
- Acquire knowledge about the scalability of different functionalities using Visual Basic for Microsoft Excel
- Utilize Macros for repetitive activities

4.1 VBA as a Programming Language

Microsoft's Visual Basic is a well-known programming language, the dialect of which is used in Microsoft Excel, mainly for creating user-defined functions. Microsoft Excel consists of a substantial list of functions that can extensively be used for performing numerous operations. However, as the nature of data varies according to the system and application, the user may want to design particular aspects of algorithms or procedures to interpret a problem. Visual Basic in Excel empowers the user to design a function as per their requirement. This is a special version of the actual programming language, solely designed to work with Microsoft Office applications, hence termed as Visual Basic for Applications (VBA). Using VBA, the user can create functions and macros for solving particular engineering or finance problems, which gives Excel the versatility to be used by the users in various domains.

4.2 VBA Interface in Excel

To code VBA in Excel, one has to have minimum knowledge on the format of Visual Basic to grasp the functions used. The code is written in a Visual Basic Editor. The editor can be opened from the "Developer" tab in the menu ribbon. In case the tab does not exist in the ribbon, do the following to enable the developer mode.

1. Right-click on the ribbon. Select "Customize the ribbon."

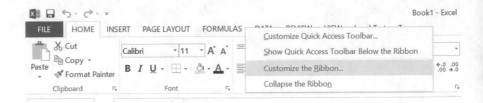

Fig. 4.1

2. Tick on the "Developer" and click "OK."

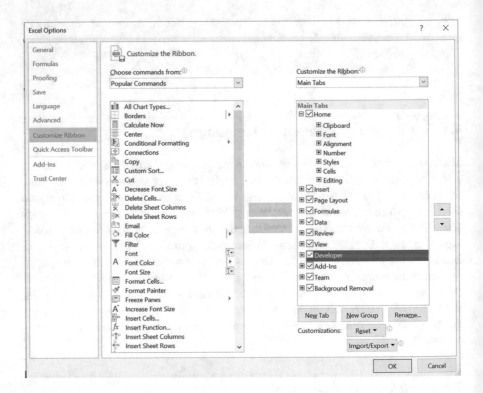

Fig. 4.2

Similarly, one can customize the tabs in the ribbon by selecting/deselecting the options from the "Main Tabs" segment of the above-shown window. For opening the Visual Basic Editor, go to "Developer > Visual Basic." A new window will appear.

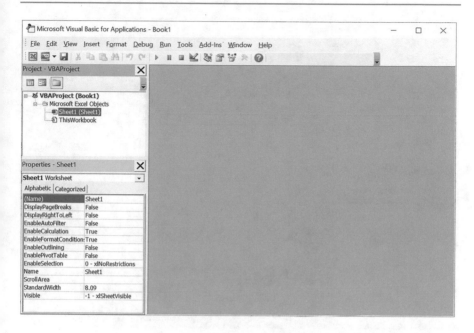

Fig. 4.3

This editor is the basic platform for coding, which includes three windows. The codes are usually written in a sheet called a Module sheet. Initially, the worksheet will not contain any Modules. To open a new module, right-click on the "VBAProject (Book1)" shown at the upper left window, and select "Insert>Module." This will create a module under the spreadsheet one is working on. Similarly, these modules can be opened, modified, and deleted by right-clicking and choosing appropriate options from them.

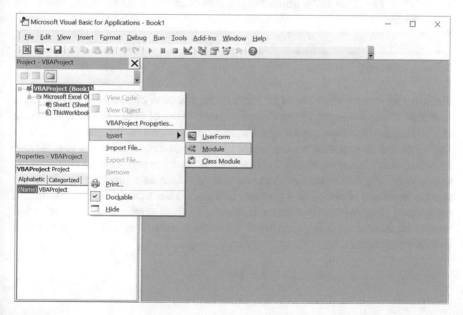

Fig. 4.4

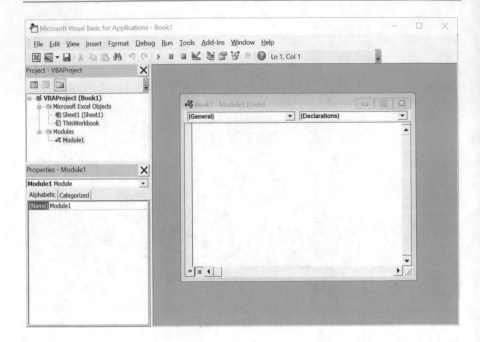

Fig. 4.5

As seen from the figure above, the interface consists of three major windows. They are:

1. Project Explorer window: This will display the hierarchy of files.
2. Property window: This displays different technical properties of the selected file.
3. Code window: This is the important segment of the VBA editor where codes can be written.

4.3   Macros

The most important application of VBA is the creation of Macros. Macros, also known as Procedures, are the module sheet consisting of codes to run a particular task in Excel. Sometimes, a set of tasks is required to be used repetitively, which costs time and energy. Macros enable the recording of the iterative task in a Macros sheet so that the Macro can be called just like a function to perform that task. Macros help to store the commands of the users and use them as a function or command using a keyboard shortcut or a couple of clicks. These Macros or Procedures are categorized into two distinct types according to their application. One is called Command macros or Sub procedures; the other is termed as Function macros or Function procedures.

Suppose one needs to create several sheets with a table of five columns with the same titles. It would take a long time to insert the table and type the attributes in each worksheet, not to mention the formatting in each table. Even if one copies and pastes the whole table, it would still be a taxing task. Macros provide the users freedom to record tasks of various aspects and assign a keyboard shortcut to operate the task as many times as possible. As the commands are recorded, the process is known as command macros or sub-procedures.

Function macros or function procedures let the users create their own function, called user-defined functions (UDF), which will work like the built-in functions. For example, one may need to convert a given set of values in inches to centimeters. In that case, the user can create a custom function named "inches2cm," the use of which will do the job just as a built-in function. The function can also be repeatedly used for other spreadsheets, enhancing the collection of functions in Excel.

Both of these Macros or procedures can be written in VBA. The basic format of writing macros will help to write and understand complicated codes later on. The basic structure of a Sub procedure includes the keyword "Sub" at the beginning of the code. The procedure ends with an "End Sub" keyword. The procedure must have a name, denoted as "ProcedureName" as follows with arguments as required. There will be empty brackets if there is no argument.

```
Sub ProcdureName(Argument1, Argument2, ...)
    VBA Statement
End Sub
```

On the other hand, a function procedure will be initialized with the keyword "Function," which will be ended with "End Function," The function named as "FunctionName" as follows may or may not have arguments. The function procedure consists of a return statement to return the result of the function.

```
Function FunctionName(Argument1, Argument2, ...)
    VBA Statement
    FunctionName = result
End Function
```

4.3.1   Sub Procedure or Command Macros

The following problem is focused on providing an insight into the coding environment for creating sub procedures and recording facility of Excel to automate the tasks.

Problem: Suppose the user wants to create a row with 5 attributes, with red bold centered font and black fill in each new sheet shown as follows.

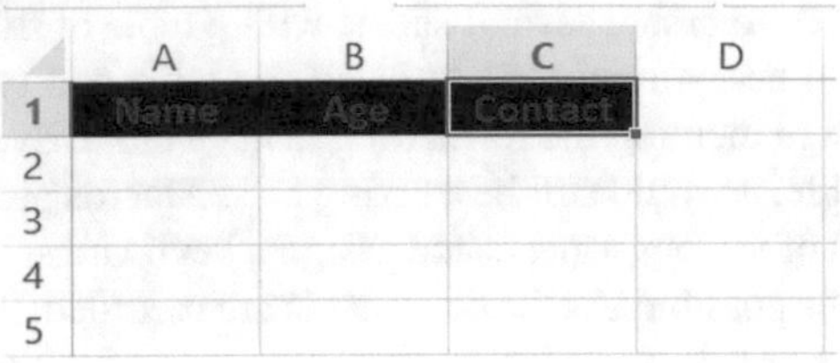

	A	B	C	D
1	Name	Age	Contact	
2				
3				
4				
5				

Fig. 4.6

Solution: Follow the steps given below:

1. As it is a sub procedure, type the basic format of the code as shown above. Let the name of the sub-procedure be "sheet_font_fill."

$$\text{Sub sheet\_font\_fill()}$$
$$|$$
$$\text{End Sub}$$

2. Plan the sequence of the task. For instance, the following could be a sequence of operations.
    (i) Create a new sheet.
    (ii) Type the attributes "Name," "Age," and "Contact" in A1, B1, and C1, respectively.
    (iii) Select A1:C1.
    (iv) Right-click and select red as the font color.
    (v) Fill with black.
    (vi) Center the writing.
    (vii) Bolden the writing.

    To create a new sheet, the following statement should be added inside the sub-procedure:

    ```
    Sheets. Add After:=ActiveSheet
    ```

According to VBA, it commands to add sheets after the active sheet. That means, if the active sheet is "Sheet 1," it will add another sheet (Sheet 2) when the code is run. Dot sign (.) in VBA simply denotes the object one wants to manipulate. In this case, as we want to manipulate the sheet by adding another one, a dot is used to indicate the "Sheets" object. Moreover, the symbol ":=" is called "Named Argument," as the argument of "adding a sheet after" is being indicated to "the active sheet." It actually passes the whole argument toward the "ActiveSheet." The code should look like this.

```
Sub sheet_font_fill()
    Sheets.Add After:=ActiveSheet
End Sub
```

3. To select cell A1 (in other words, row 1 and cell 1) and type "Name," the following statement is written:

   ```
   Range("A1").Select
   ActiveCell.FormulaR1C1 = "Name"
   ```

   Here, 'Range("A1")' indicates the chosen cell. "Select" is the function to describe the selection of the object 'Range("A1")'. In the "ActiveCell" previously mentioned, the provided text is assigned to row 1 and cell 1 (denoted by "R1C1"). "FormulaR1C1" is a function to manipulate the active cell with the text mentioned following the assignment symbol "=." Similar statements should be written for "Age" and "Contact." The code at the end of this stage should look like this.

   ```
   Sub sheet_font_fill()
       Sheets.Add After:=ActiveSheet
       Range("A1").Select
       ActiveCell.FormulaR1C1 = "Name"
       Range("B1").Select
       ActiveCell.FormulaR1C1 = "Age"
       Range("C1").Select
       ActiveCell.FormulaR1C1 = "Contact"
       Range("A1:C1").Select
   End Sub
   ```

4. The next task will be to select A1:C1 for formatting. The statement for selection will be:

   ```
   Range("A1:C1").Select
   ```

5. With this particular selection, further operations can be performed. A new nested section with the keyword "With" will be used, which will continue some tasks as long as a particular condition is met. In this case, as long as the cell range is selected, the fonts can be manipulated. The segment is ended with an "End With" keyword at the end.

   To select red as the font color, the following lines are typed:

   ```
   With Selection.Font
           .Color = RGB(255, 0, 0)
   End With
   ```

   Here, while the "Selection.Font" condition is met, the function "Color" can be assigned to the "Font." According to the RGB (Red, Green, Blue) format of numerating colors, red colors are denoted as (255,0,0), hence the second line of the code. Finally, the segment for coloring the font while the selection is facilitated ends with "End With." The code looks like this:

```
Sub sheet_font_fill()
    Sheets.Add After:=ActiveSheet
    Range("A1").Select
    ActiveCell.FormulaR1C1 = "Name"
    Range("B1").Select
    ActiveCell.FormulaR1C1 = "Age"
    Range("C1").Select
    ActiveCell.FormulaR1C1 = "Contact"
    Range("A1:C1").Select
    With Selection.Font
        .Color = RGB(255, 0, 0)
    End With
End Sub
```

6. For filling the selected cells with black colors, the following statements are used.

```
With Selection.Interior
    .ColorIndex = 1
End With
```

In this case, the interior of the cell is marked by the function "Interior." The first line means that with the interior being selected, the following codes should be implemented. In the next line, 1 is assigned as the color index for the black color. The number is according to the 56-color format of Excel VBA, where 1 is considered as black, 0 as white, 56 as gray, and so on. The segment, like the previous step, is ended with "End With."

```
Sub sheet_font_fill()
    Sheets.Add After:=ActiveSheet
    Range("A1").Select
    ActiveCell.FormulaR1C1 = "Name"
    Range("B1").Select
    ActiveCell.FormulaR1C1 = "Age"
    Range("C1").Select
    ActiveCell.FormulaR1C1 = "Contact"
    Range("A1:C1").Select
    With Selection.Font
        .Color = RGB(255, 0, 0)
    End With
    With Selection.Interior
        .ColorIndex = 1
    End With
End Sub
```

7. To center the texts, the "HorizontalAlignment" function of the selected portion as a whole should be centered. The keyword for centering in VBA is "xlCenter."

```
With Selection
    .HorizontalAlignment = xlCenter
End With
```

Moreover, to bolden the selected font, the following statement should suffice.

```
Selection.Font.Bold = True
```

Thus, the complete code becomes:

```
Sub sheet_font_fill()
    Sheets.Add After:=ActiveSheet
    Range("A1").Select
    ActiveCell.FormulaR1C1 = "Name"
    Range("B1").Select
    ActiveCell.FormulaR1C1 = "Age"
    Range("C1").Select
    ActiveCell.FormulaR1C1 = "Contact"
    Range("A1:C1").Select
    With Selection.Font
        .Color = RGB(255, 0, 0)
    End With
    With Selection.Interior
        .ColorIndex = 1
    End With
    With Selection
        .HorizontalAlignment = xlCenter
    End With
    Selection.Font.Bold = True
End Sub
```

After writing the code, it should be checked for possible errors and run to execute if no error is found. Click on the following button to run the code and check the outcomes in the Excel file.

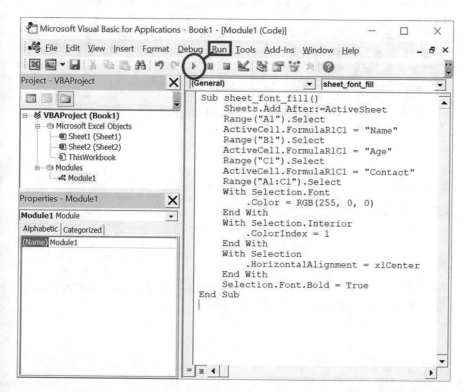

Fig. 4.7

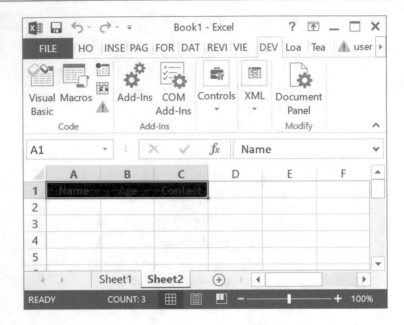

Fig. 4.8

It is seen that a new sheet has been created with the exact feature mentioned in the problem. A similar event will occur just by clicking the "Run" symbol in the editor.

There is an alternate method to create such a macro without using VBA. Excel provides a great tool for recording the real-time operation, which is converted into the corresponding code to be stored as a macro. The option for recording the macro is available on the "Developer" tab and is described as follows:

1. From the "Developer" tab, click on "Record Macro."

Fig. 4.9

2. A window will appear to configure the macro. Type the macro name, shortcut key, and description as follows. Click "OK."

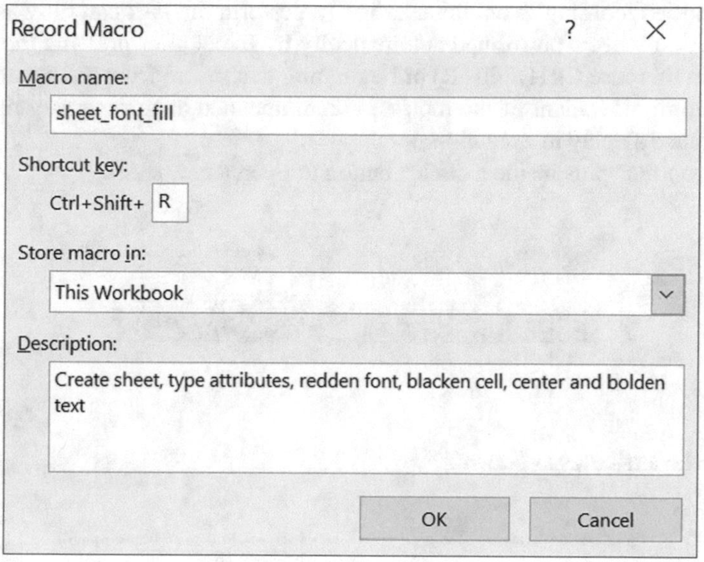

Record Macro   ?   ×

Macro name:

sheet_font_fill

Shortcut key:

Ctrl+Shift+ R

Store macro in:

This Workbook   ∨

Description:

Create sheet, type attributes, redden font, blacken cell, center and bolden
text

                                        OK   Cancel

Fig. 4.10

3. Once the recording starts, an option to "Stop Recording" appears instead of
"Record Macro."

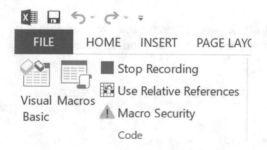

Fig. 4.11

The bottom symbol for creating macro also changes into a "Stop" symbol.

Fig. 4.12

4. While the recording is on, the user has to perform the task carefully as each of the action will be performed automatically by Excel after pressing the shortcut key (in this case Ctrl+Shift+R) or by running that macro from the list of macros. Minimum movement of the mouse is recommended as unnecessary movement will cause a delay in execution.
5. Click on the "plus inside a circle" button to open a new sheet.

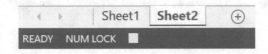

Fig. 4.13

6. Type the attributes as shown.

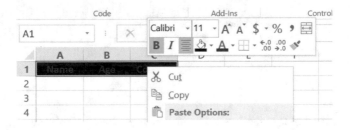

Fig. 4.14

7. Select cell A1:C1. Right-click on the selection and redden the font, fill the cells with black color, centralize the texts and bolden them.

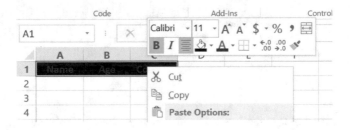

Fig. 4.15

8. All the tasks being completed, stop the recording either by clicking on "Stop Recording" at the ribbon tab or by selecting the "Stop" symbol at the status bar.
9. The record has been converted into VBA code, and the selected keyboard shortcut has been assigned to repeat the whole task. By typing the keyboard shortcut or by running the macro, Excel will automatically operate the VBA code automatically, regenerating the complete task in less than a second.

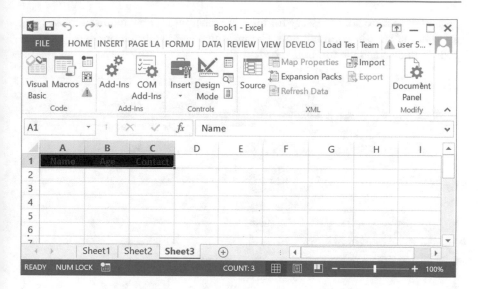

Fig. 4.16

The created code can be seen by clicking on the "Macros" in the "Developer" tab. After the appearance of the window, select "Step Into" to jump into the code.

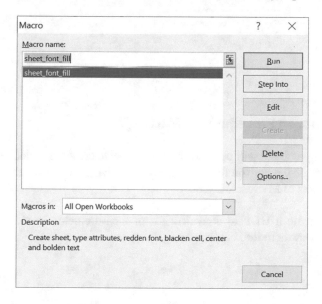

Fig. 4.17

As seen from the code, it consists of some additional information apart from the basic code we have written before. It is because while configuring certain parameters like centering or filling cells, Excel takes into consideration many internal properties that are not necessary for this particular purpose but required for keeping up with the built-in configuration. Compare this code with the previously written code to understand the difference better.

Fig. 4.18

4.3.2   Function Procedures or Macros

The following problem will provide an idea on how to create user-defined functions (UDF) alongside the built-in functions to augment the list and take advantage of the protean features of Microsoft Excel.

Problem: Create a UDF to convert a list of measurements in inches as shown below to convert them into centimeters.

	A	B
1	1	
2	42	
3	24	
4	64	
5	27.5	
6	87	
7	102	
8	93	
9		

Fig. 4.19

Solution: 1. To create a function procedure, write the basic structure of the code in a VBA module as follows. Name the function "inches2cm(inches)." The argument "inches" in the function determines the value it will take to process.

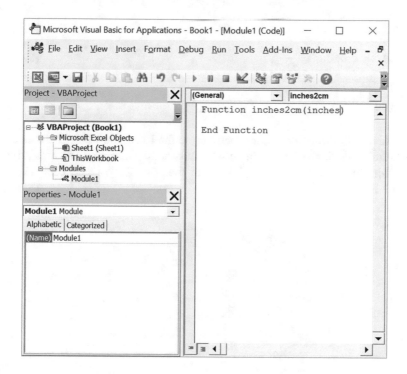

Fig. 4.20

2. As 1 inch = 2.54 cm, write the relationship in between the function and assign the
   result to the function name.

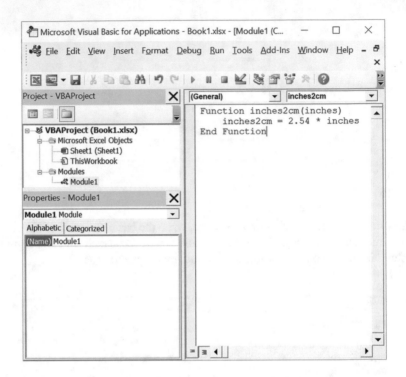

Fig. 4.21

3. On saving the workbook, after returning to the Excel file where the data exists,
   this function "inches2cm" can be used in exactly the same manner as a typical
   built-in function. Select cell A2, write the formula "=inches2cm(A1)," and
   press Enter.

	A	B	C
1	1	=inches2cm(A1)	
2	42		
3	24		
4	64		
5	27.5		
6	87		
7	102		
8	93		
9			

Fig. 4.22

4. On getting the value, select the cell again. Click on the tiny solid square at the down-right corner of the cell. Then, pull the formula downward for consecutive values. Excel will automatically formulate and display that the function is working.

	A	B	C
1	1	2.54	
2	42	106.68	
3	24	60.96	
4	64	162.56	
5	27.5	69.85	
6	87	220.98	
7	102	259.08	
8	93	236.22	
9			
10			

Fig. 4.23

The name of the function should be precise but should be able to mention the purpose with the word, so the name should be picked carefully. As the name of the function will be used repeatedly, it should be short but self-explanatory.

4.3.3 Executing a VBA Sub Procedure Using a Button

We can assign a macro to specific objects that are allowed to insert in an Excel worksheet. These objects can be classified into three [1], namely form controls, activeX controls, and inserted objects such as clip art, shapes, WordArt, SmartArt, text box, charts, pictures, etc. The steps to attach a macro to a Form control button are delineated below:

Step #1: Inserting a Button
From the ribbon, go to the Developer tab. Click on Insert > Button Form Control.

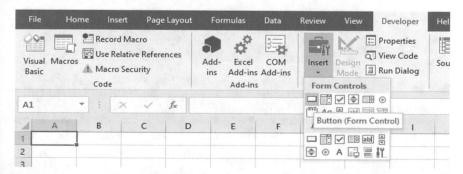

Fig. 4.24

Step #2: Creating the Button

Once the Button Form Control is clicked, the button can be created by clicking on the top-left corner of the cell where the button is desired. For instance, if the button is desired in the cell C2, click on the top-left corner of that cell.

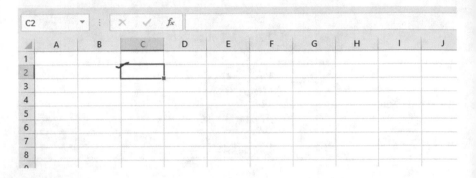

Fig. 4.25

Step #3: Assigning a Macro to the Button

After clicking on the desired cell, the Assign Macro dialog box appears.

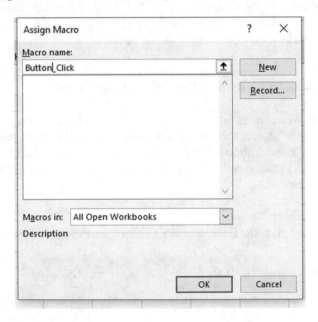

Fig. 4.26

Excel suggests a macro name according to the name of the button. Here, this is "Button_Click". But the suggested macro may not match the requirements. So, select the desired macro, and click on OK. In the example discussed throughout this chapter, a suitable button name is Delete Blank Rows.

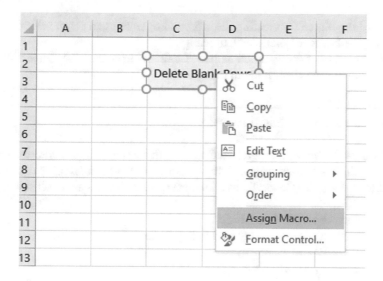

Fig. 4.27

Assigned macro: The VBA sub-procedure associated with a button can be changed by right-click > "Assign Macro…".

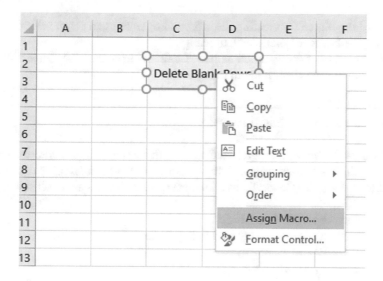

Fig. 4.28

The Assign Macro dialog box will reappear. We can select the particular VBA Sub procedure assigned to the button.

Exercise 4.1:

1. Create a sub-procedure using VBA code to copy the following set in a new sheet.

	A	B	C	D	E	
1	Sl. No	Team	Judge A	Judge B	Judge C	
2	1	Alpha				
3	2	Beta				
4	3	Gamma				
5	4	Zeta				
6	5	Phi				
7	6	Sigma				
8	7	Omega				
9	8	Eta				
10	9	Nabla				
11	10	Delta				
12						
13						

Fig. 4.29

2. Record the Marco for creating the aforesaid operations so that it can be used by typing the keyboard shortcut "Ctrl+Shift+T".
3. Create a user-defined function named "Cen2Fer" to convert the temperature of a certain area from Centigrade to Fahrenheit.

4.4 Creating an Excel Add-in

An add-in is defined as a program attached to Excel that provides additional functionality. Three steps are executed for creating an add-in in Excel and making it available in the QAT (Quick Access Toolbar) [2]. These steps are described below.

Write/Record the Code in a Module
A simple code can be used to highlight all cells having error values:

```
Sub HighlightErrors()
Selection.SpecialCells(xlCellTypeFormulas, xlErrors).Select
    Selection.Interior.Color = vbRed
End Sub
```

Right-click on any object in the workbook. Click on Insert > Module for inserting a module object.

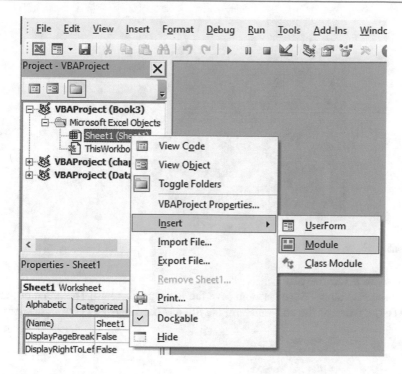

Fig. 4.30

On the module, double-click, and then copy the given code and paste it there.

Fig. 4.31

Go back to the worksheet by pressing Alt+F11.

Fig. 4.32

Save as an Excel Add-in

Go to the File tab > "Save As" and save the file type as .xlam. The name of the file will be the name of the add-in. Here, the file has been saved as Highlight Errors.

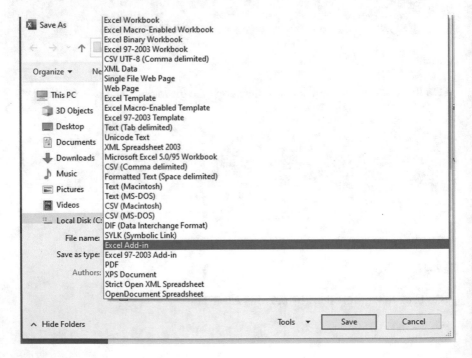

Fig. 4.33

In a fresh workbook, go to the Developer tab and select Add-ins > Excel Add-ins.

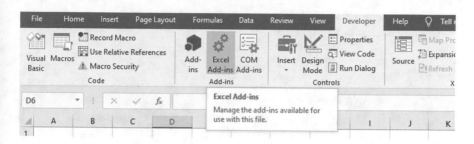

Fig. 4.34

Find and select the saved file from the Add-ins dialogue box. Click OK.

Fig. 4.35

Add the Macro to the QAT
The add-in is now activated. By adding the macro to the QAT, it can be run with a single click. Save and Install the Add-in, for which:

- Right-click anywhere on the ribbon and click on Customize Quick Access Toolbar.
- Open the dialog box named Excel Options. Then, select Macros from the drop-down menu below "Choose commands from."

- Select the macro "HighlightErrors" and click on Add>>. Thus, the macro will be added to the list to the right.

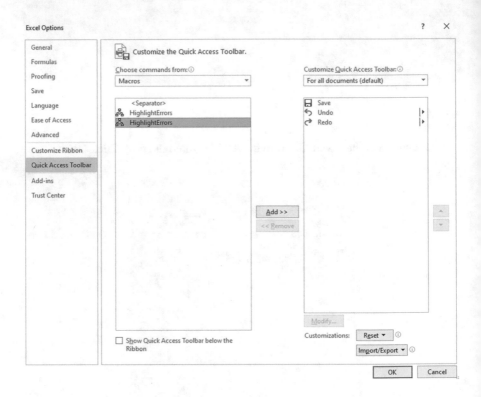

Fig. 4.36

Click OK. Thus the macro will be added.

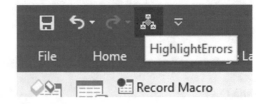

Fig. 4.37

You can select any dataset from any workbook and click on the macro icon in the QAT for running the code.

4.5   Conclusion

This chapter is aimed at explaining the concept of Visual Basic for Application (VBA) in Microsoft Excel. VBA is a feature that can enable users to create new functions according to their needs. VBA thus makes Excel very user-oriented and caters to all users. The chapter explains macros and add-ins and elaborates on how to create and use them. This chapter is useful for anyone who wishes to master MS Excel and be able to exploit the full potential of this amazing tool.

References

1. https://powerspreadsheets.com/vba-sub-procedures/
2. https://trumpexcel.com/excel-add-in/

MS Excel in Engineering Data

5

Learning Objectives

- Manage files and data using Microsoft Excel
- Import data from various sources in Excel
- Export prepared data using Microsoft Excel
- Learn the basics of numerical analysis in Microsoft Excel
- Create and perform basic operations on Matrices in Microsoft Excel
- Plot 2D or 3D equations in Microsoft Excel
- Interpolate or extrapolate data using trendline

5.1   File Management in Excel

Microsoft Excel allows data from an external source to be used in the spreadsheet itself. Even though data can be copied and pasted, and multiple sheets can be created to work with the data. But it is possible that the amount of data is so enormous that it is saved in another file or system. It is also probable that data has been accumulated from another system in a file format that is not .csv. For such cases, Microsoft Excel has a feature to get the file from an external source. This system of acquiring files from an external source is known as "Import" in Excel. Sometimes, after collecting data in Excel, it may be necessary to convert it to another file format, or "Export" the file, as it is called in Excel so that it can be used in other systems. As data can be stored in other file formats such as .csv, .txt, etc., exporting Excel files enable the file to be converted according to the format. Such format conversions are of great importance while focusing particularly on data while working instead of software, enabling interoperability among miscellaneous systems.

5.1.1   Import Files to Excel

Excel permits importing files of formats, usually of .csv or .txt. The simplest way to import a file is to select "File>Open>Browse." Select the extension of the file from the drop-down menu. Then click the required file from the sorted list. Importing a .csv file is straightforward. But in order to import a .txt file, the users should use "Excel Import Wizard," where options are available to import data from Microsoft Access databases, websites, text files, servers, or external connections. For instance, for importing data from a .txt file having random data, as shown below, go to "Data > Get External Data > From Text" to import the data into the existing workbook.

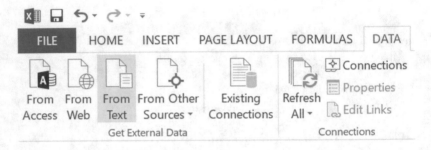

Fig. 5.1

A window for browsing the text file will appear. Select the desired text file consisting of the data. The demo data is as follows:

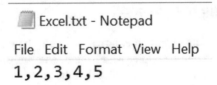

Fig. 5.2

On selecting the file, the following screen will appear:

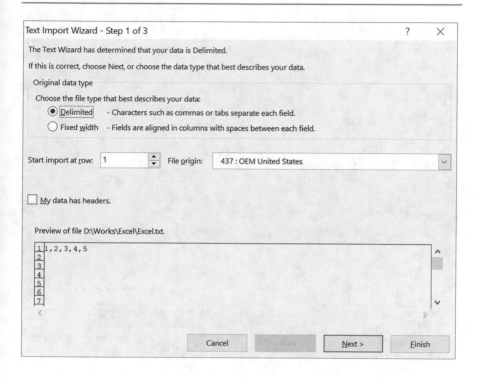

Fig. 5.3

It is seen that the data is separated using a delimiter. So, the file type should be selected as mentioned above. The delimiter is the symbol that separates the data elements from each other. It may be a comma, semicolon, alphabets, or even space. After clicking "Next," the delimiter is to be selected.

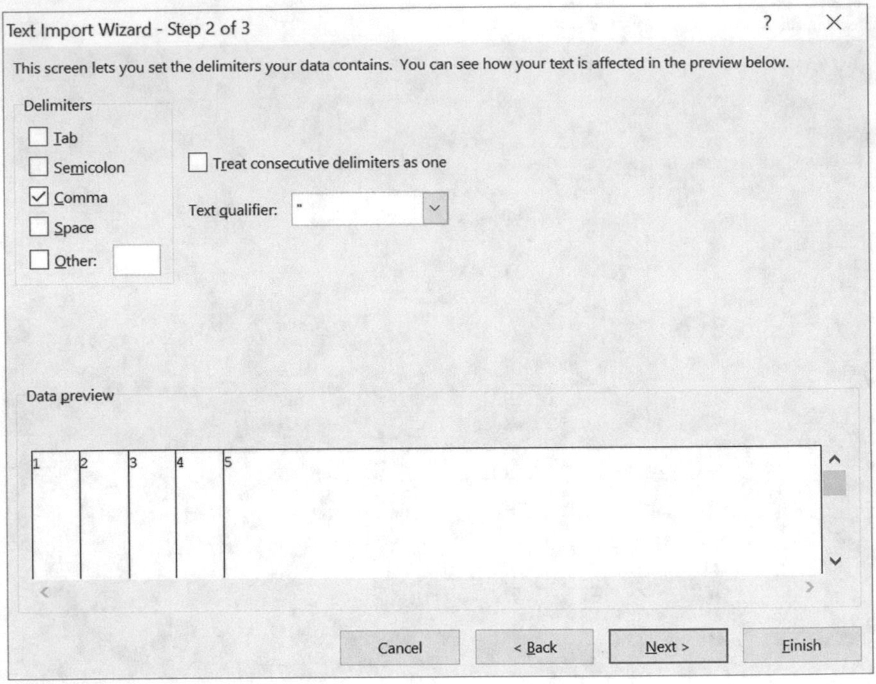

Fig. 5.4

Select "Comma" and see how the data preview has changed. It means that the data will be represented in a row. Click "Next" for the final step.

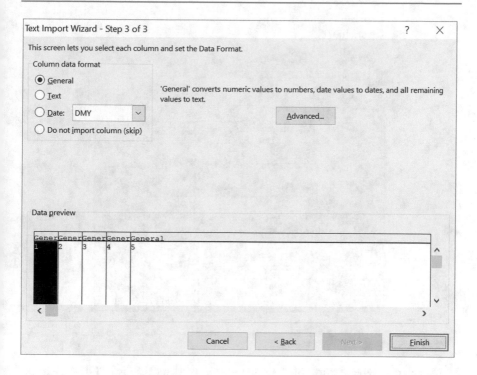

Fig. 5.5

This window is for individual manipulation of the column. Click "General" and then "Finish."

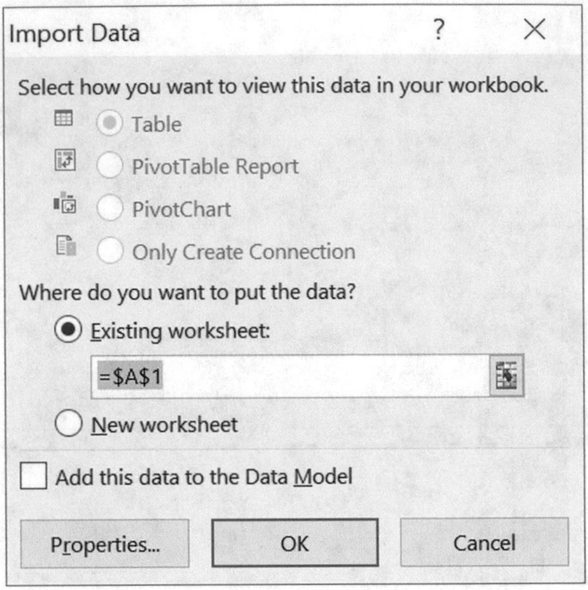

Fig. 5.6

On selecting the starting of the column, the worksheet, click "OK." Then the imported data will be available as follows:

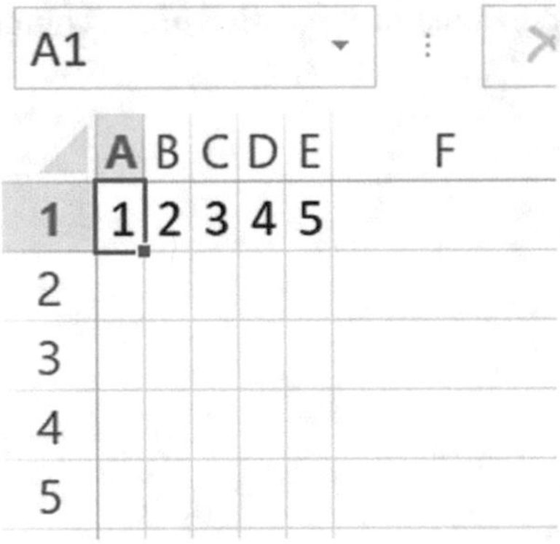

Fig. 5.7

In case one wants to transpose the data, one can use the TRANSPOSE function. Another way of transposing the data is using the "Paste Special" as follows:

1. Select the data and copy it.
2. Select a column with the same dimension for placing the elements.

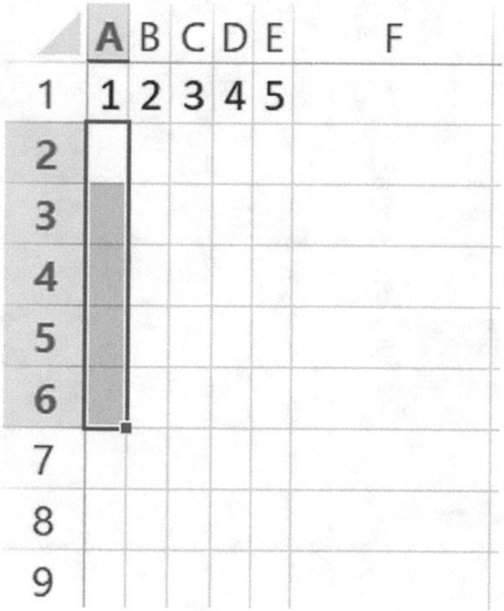

Fig. 5.8

3. Right-click on the selected range of cells and click "Paste Special." Then the following window will appear.

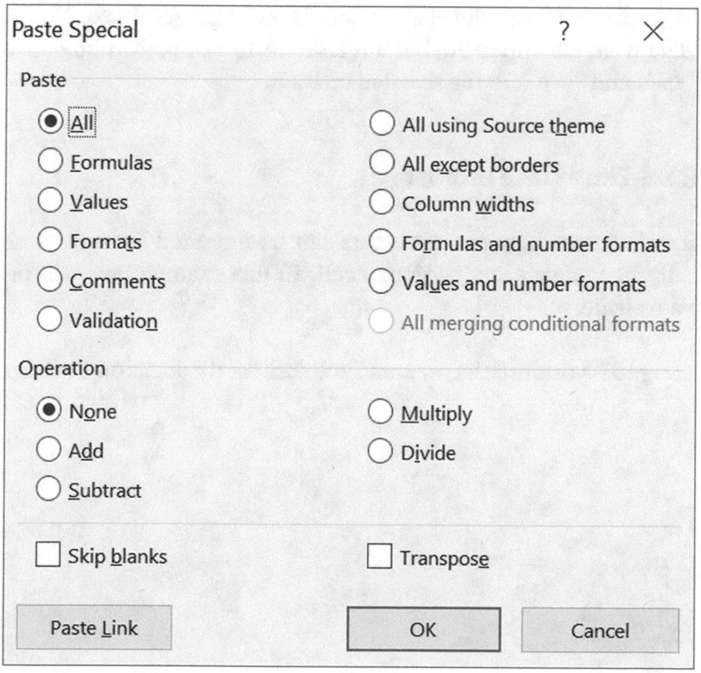

Fig. 5.9

4. Tick on "Transpose" and press "OK."

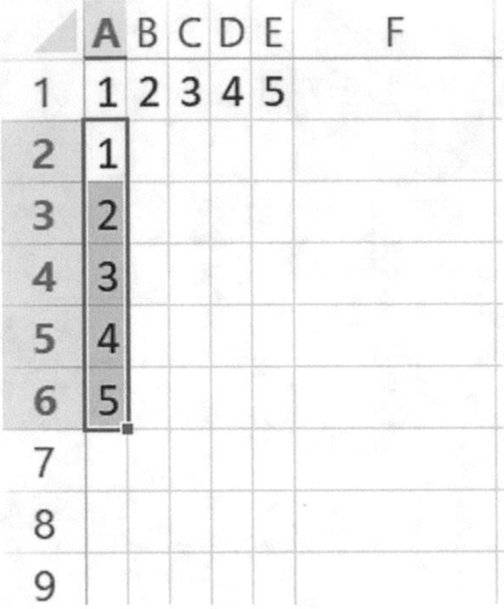

Fig. 5.10

5. Finally, delete the initial row and format it accordingly.

As seen from the "Paste Special" option, it has a range of use. Not only it can paste the data from the clipboard, but it is also capable of performing operations on the stored data and save it to the selected cells.

5.1.2   Real-Time Data Import

This section discusses how real-time data can be imported from the web and can automatically be updated in a particular cell. In this example, we will obtain time data from a website.

1. Navigating to "worldtimeserver.com," we surf for the location in Oregon.

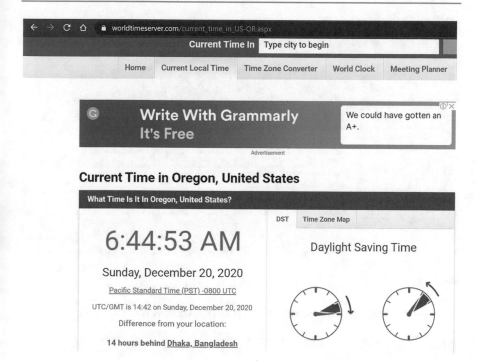

Fig. 5.11

2. Go to Excel, and select Ribbon > Data.

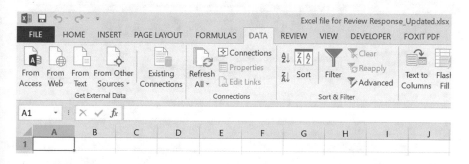

Fig. 5.12

3. Click on "From Web," a new tab will appear.

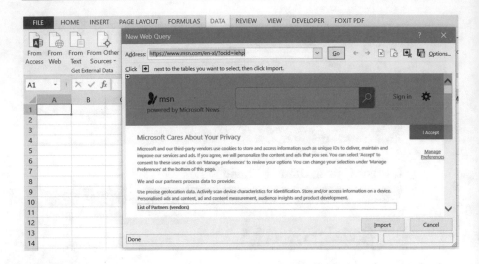

Fig. 5.13

4. Copy the link from the website. Paste it in the Address bar. Press "Go."

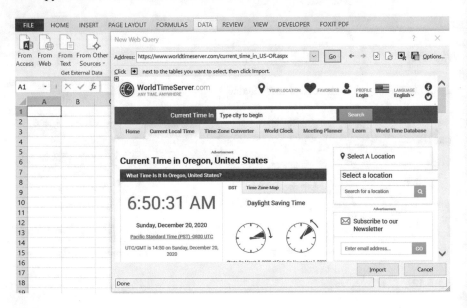

Fig. 5.14

5. Click "Import." A new window appears after connecting to the server.

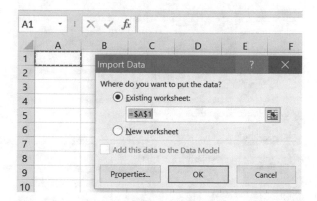

Fig. 5.15

6. As we want to display the data in the existing server, click OK. Many fields from the website will appear in the existing worksheet.

Fig. 5.16

After navigating the column, we can see the time data in cell A31. This is the time data that will be automatically updated from the website after pressing "Refresh All" in the "Data" Tab.

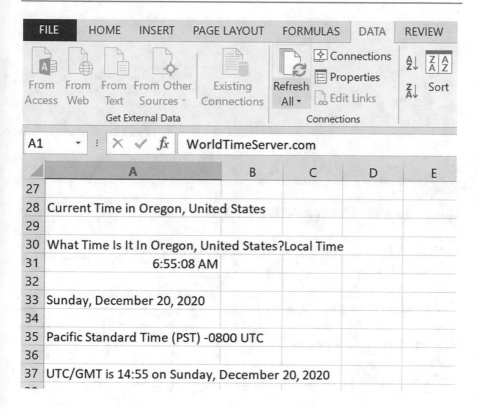

Fig. 5.17

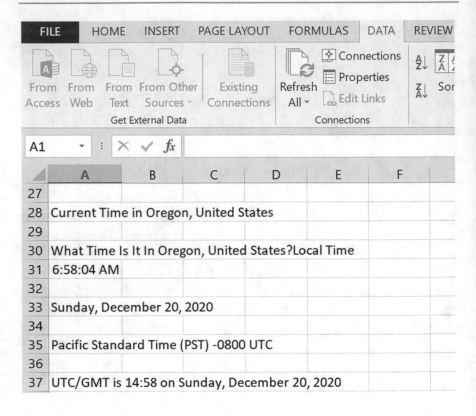

Fig. 5.18

7. To automate the process, go to the drop-down menu in "Refresh All" and click "Connection Properties."

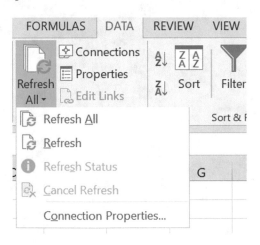

Fig. 5.19

8. Check the box beside "Refresh every." Then provide the desired time in minutes to update automatically.

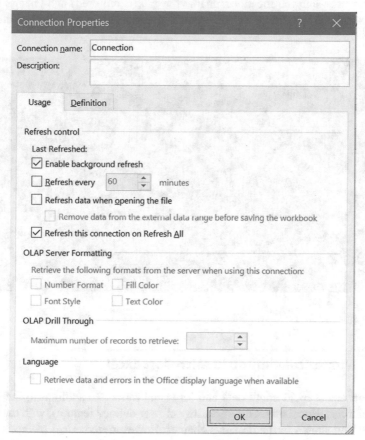

Fig. 5.20

After clicking OK, the real-time import of time data can be noticed.

This is a very important feature for engineering that can also be used for real-time weather monitoring, stock price analysis, and other applications requiring automation.

5.1.3   Export Files from Excel

To save the Excel data in other formats, after completing the work, select "Files>Save as." After browsing the location, rename the file (if required) and change the file type (extension) according to the system requirement. On selecting the type, click "Save."

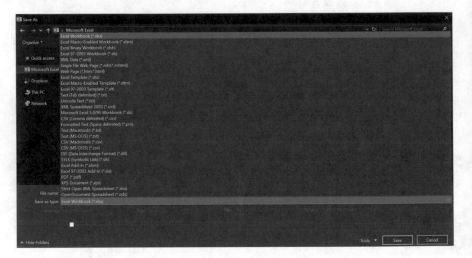

Fig. 5.21

5.2   Numerical Method in Microsoft Excel

Microsoft Excel is all about data and its visualization. In addition to the built-in functionalities, Excel offers the versatility of user-defined features, with the help of VBA coding shown in the previous chapter. However, numerical methods are the basis for most of the mathematical operations. Interpolation–extrapolation, integration–differentiation, linear and differential equations, regressions, and curve fitting are some of the prominent topics in the study of numerical methods. Excel contains some prepared functions for such operations, but when it comes to tabulate values and check certain conditions to iterate a process, programming can be of great help, enabling freedom to choose the conditions, process, and parameters according to the requirement. Coding knowledge in VBA will help the users to a large extent to deal with numerical problems to program the mathematical procedures. This chapter will particularly deal with matrices, linear equations, and curve fitting algorithms to help the readers to grasp the core of numerical methods and to make understand how Microsoft Excel can prove to be an excellent tool to work with similar problems.

5.2.1   Gauss-Seidel Method

Three matrices—labeled A, B, and C—are provided, as denoted in Fig. 5.22.

	A	B	C	D	E
1	Given:				
2		X	Y	Z	W
3	A:	4	22	-13	-128
4	B:	15	-13	4	111
5	C:	8	8	17	10

Fig. 5.22

Rearrange the values:

7	Arranged according to domminance (Strictly Diagonally Domminant)				
8		X	Y	Z	W
9	A:	19	-13	4	111
10	B:	4	22	-13	-128
11	C:	8	8	17	10
12					

Fig. 5.23

Build the equations:

Equations:		Isolation:	
19X-13Y+4Z=111		X=(111+13Y-4Z)/19	
4X+22Y-13Z=-128		Y=(-128-4X+13Z)/22	
8X+8Y+17Z=10		Z=(10-8X-8Y)/17	

Fig. 5.24

Cell preparation for iterations:

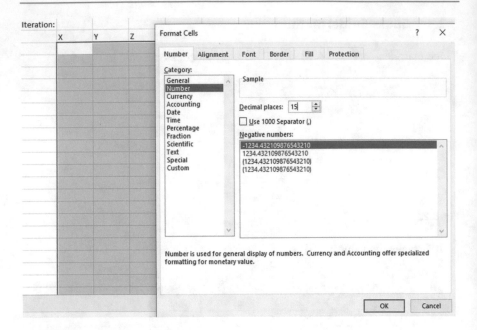

Fig. 5.25

First iteration for *x*:

Iteration:			
	X	Y	Z
0	0.000000000000000	0.000000000000000	0.000000000000000
1	=(111+13*J3-4*K3)/19		

Fig. 5.26

First iteration for *y*:

Iteration:			
	X	Y	Z
0	0.000000000000000	0.000000000000000	0.000000000000000
1	5.842105263157890	=(-128-4*I3+13*K3)/22	

Fig. 5.27

First iteration for *z*:

Iteration:			
	X	Y	Z
0	0.000000000000000	0.000000000000000	0.000000000000000
1	5.842105263157890	-5.818181818181820	=(10-8*I3-8*J3)/17
–			

Fig. 5.28

Result of the first iteration:

Iteration:			
	X	Y	Z
0	0.000000000000000	0.000000000000000	0.000000000000000
1	5.842105263157890	-5.818181818181820	0.588235294117647

Fig. 5.29

	A	B	C	D	E	F	G	H	I	J	K
1	Given:							Iteration:			
2		X	Y	Z	W				X	Y	Z
3	A:	4	22	-13	-128			0	0.000000000000000	0.000000000000000	0.000000000000000
4	B:	19	-13	4	111			1	5.842105263157890	-5.818181818181820	0.588235294117647
5	C:	8	8	17	10			2	1.737405009850830	-6.532789192231920	0.576977202364199
6								3	1.250833246922540	-5.793132564030400	2.844886674061690
7	Arranged according to domminance (Strictly Diagonally Domminant)							4	1.279459472176740	-4.364536646585830	2.725787913933110
8		X	Y	Z	W			5	2.281993259929040	-4.440117954889840	2.040036317368980
9	A:	19	-13	4	111			6	2.374648490366110	-5.027613677905430	1.603823385863910
10	B:	4	22	-13	-128			7	2.064512033882830	-5.302222270237890	1.836689500018500
11	C:	8	8	17	10			8	1.827597499307020	-5.108231119785950	2.111863640637680
12								9	1.902396888433260	-4.902552848588100	2.132062880225380
13	Equations:			Isolation:				10	2.038871655129110	-4.904216823218330	2.000073393014040
14		19X-13Y+4Z=111			X=(111+13Y-4Z)/19			11	2.065520406637140	-5.007024205060630	1.936633020277280
15		4X+22Y-13Z=-128			Y=(-128-4X+13Z)/22			12	2.008534381742250	-5.049356925588360	1.972472375728700
16		8X+8Y+17Z=10			Z=(10-8X-8Y)/17			13	1.972024761286130	-5.017818029204360	2.019210608868760
17								14	1.983764378150960	-4.983561869538660	2.021549773137990
18								15	2.006710347549760	-4.984314111900460	1.999904701829510
19								16	2.010752512525060	-5.001276375746160	1.989460594988560
20								17	2.001345512386620	-5.008182832511310	1.995540641515810
21								18	1.995340032173140	-5.002879714083680	2.003217562411620
22								19	1.997352287750830	-4.997251446242790	2.003548085604960
								20	2.001123624922210	-4.997422991722580	1.999052545172690

Fig. 5.30

Find where the value starts repeating.

H	I	J	K
85	2.000000000000100	-4.999999999999840	1.999999999999970
86	2.000000000000120	-5.000000000000040	1.999999999999880
87	2.000000000000000	-5.000000000000090	1.999999999999960
88	1.999999999999940	-5.000000000000020	2.000000000000040
89	1.999999999999980	-4.999999999999960	2.000000000000040
90	2.000000000000020	-4.999999999999970	1.999999999999990
91	2.000000000000020	-5.000000000000010	1.999999999999980
92	2.000000000000000	-5.000000000000020	1.999999999999990
93	1.999999999999990	-5.000000000000000	2.000000000000010
94	2.000000000000000	-4.999999999999990	2.000000000000010
95	2.000000000000000	-5.000000000000000	2.000000000000000
96	2.000000000000000	-5.000000000000000	2.000000000000000
97	2.000000000000000	-5.000000000000000	2.000000000000000
98	2.000000000000000	-5.000000000000000	2.000000000000000
99	2.000000000000000	-5.000000000000000	2.000000000000000
100	2.000000000000000	-5.000000000000000	2.000000000000000
101	2.000000000000000	-5.000000000000000	2.000000000000000
102	2.000000000000000	-5.000000000000000	2.000000000000000
103	2.000000000000000	-5.000000000000000	2.000000000000000
104	2.000000000000000	-5.000000000000000	2.000000000000000
105	2.000000000000000	-5.000000000000000	2.000000000000000
106	2.000000000000000	-5.000000000000000	2.000000000000000
107	2.000000000000000	-5.000000000000000	2.000000000000000

Fig. 5.31

This is the solution: $X = 2$, $Y = -5$, $Z = 2$.

Exercise 5.1: Find the solutions using the Gauss–Seidel method in Excel:

$$4x_1 - x_2 - x_3 = 3$$
$$-2x_1 + 6x_2 + x_3 = 9.$$
$$-x_1 + x_2 + 7x_3 = -6$$

5.3   Matrices

Matrix operation in MS Excel is pretty straightforward, as each element can be put in a cluster of cells, which can be annotated as a matrix per se. For example, to save the following matrix in Excel, do the following:

$$A = \begin{bmatrix} 3 & 7 & -2 \\ 1 & 3 & 5 \\ 9 & -4 & 4 \end{bmatrix}.$$

1. Select a cell of the same dimension as the matrix (in this case, 3R by 3C). Right-click on the cell and select "Define Name."

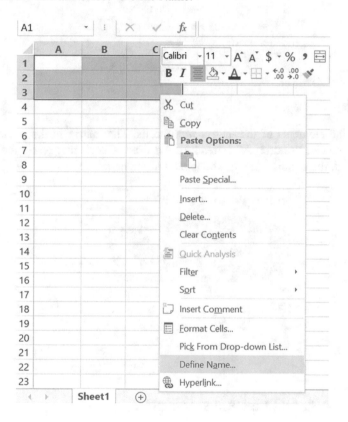

Fig. 5.32

2. Name the matrix "Matrix_A." The name should not contain any space in between or should not match with any built-in objects. Click "OK."

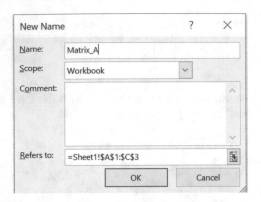

Fig. 5.33

3. Type the elements of the matrix in the cells. The matrix in the particular cell range named "Matrix_A" is ready to be used. Check that while the range is selected, the name of the matrix appears at the name tab.

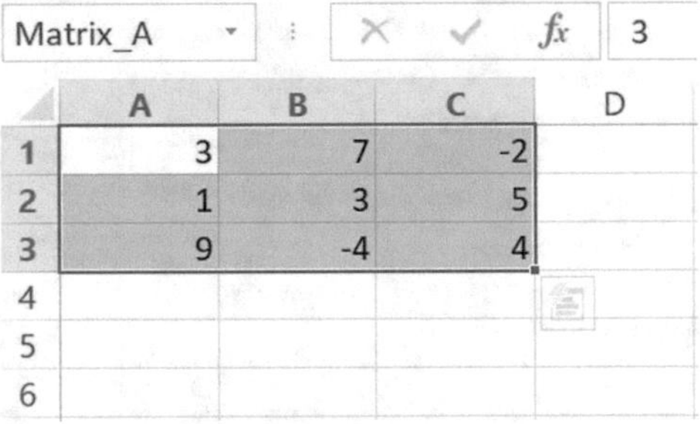

Fig. 5.34

5.3.1   Matrix Addition and Subtraction

To add 6 with Matrix_A, select another set of cells with the same dimension so that the result can fit the order. Type the formula "Matrix_A+6" in the formula bar.

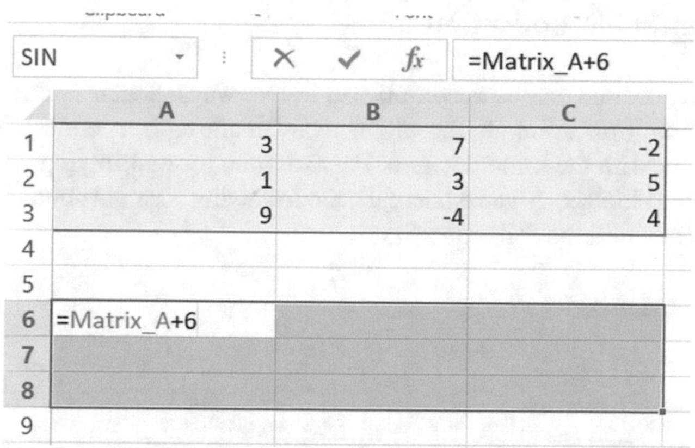

Fig. 5.35

Press "Ctrl+Shift+Enter" on the keyboard to display the output. "Ctrl+Shift+Enter" in this context is required for matrix operation. Pressing otherwise would indicate an error in operation. It is seen that after clicking on the resultant matrix, the formula appears as {=Matrix_A+6}. It is because in Excel, { } indicates a matrix array.

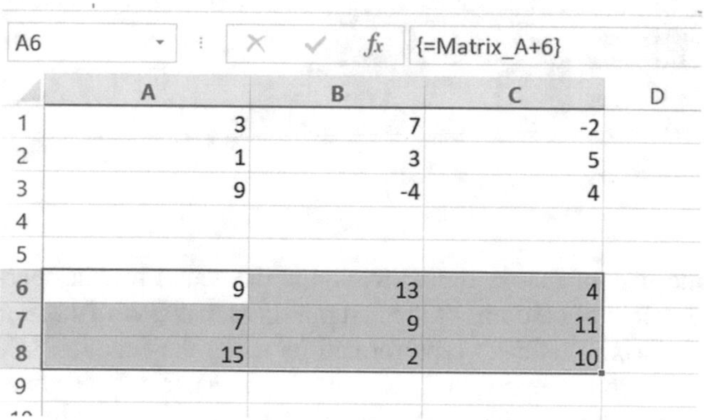

Fig. 5.36

For subtraction of 6 from Matrix_A, the formula should be {=Matrix_A-6}.

Exercise 5.2: Add and subtract these two matrices and define their name as Matrix_C and Matrix_D, respectively.

$$A = \begin{bmatrix} 5 & 2 & -2 \\ 7 & 0 & 2 \\ 6 & 1 & 23 \end{bmatrix} \text{ and } B = \begin{bmatrix} -5 & 1 & 9 \\ 0 & -8 & 2 \\ 4 & 5 & 3 \end{bmatrix}.$$

5.3.2   Matrix Multiplication

Matrix multiplication can be accomplished in two ways. Scalar multiplication will lead to the multiplication of the elements with the corresponding elements of another matrix with the same position. For example, for multiplying corresponding of two matrices Matrix_A and Matrix_B, the following formula should suffice. The example is shown in the following figure.

{=Matrix_A*Matrix_B}

A9			✕	✓	*fx*	{=Matrix_A*Matrix_B}	
	A	**B**	**C**	D	E	F	
1	1	2	3				
2	-4	0	-6				
3	7	-8	9				
4							
5	-9	8	-7				
6	-4	5	0				
7	3	2	1				
8							
9	-9	16	-21				
10	16	0	0				
11	21	-16	9				
12							

Fig. 5.37

The second way of matrix multiplication is the actual matrix operation, which requires the number of columns of the first matrix to match with the number of rows of the other matrix. This can be performed by using the function MMULT(A,B), where A and B are the matrices. If Matrix_A and Matrix_B are two matrices as shown in the previous example to be multiplied, the function is to be used instead of the formula previously shown.

A9			✕	✓	*fx*	{=MMULT(Matrix_A,Matrix_B)}	
	A	**B**	**C**	**D**	**E**	**F**	**G**
1	1	2	3				
2	-4	0	-6				
3	7	-8	9				
4							
5	-9	8	-7				
6	-4	5	0				
7	3	2	1				
8							
9	-8	24	-4				
10	18	-44	22				
11	-4	34	-40				
12							

Fig. 5.38

Compare two results in this section. The first multiplication is the scalar multiplication, performing element-by-element multiplication. The second one is the actual matrix multiplication maintaining the rules of matrices. Note that while performing matrix operation in Excel, after writing the formula in the selected cells, "Ctrl+Shift+Enter" must be pressed as the key combination are assigned for matrix operations only.

Exercise 5.3: For the Matrix_A and Matrix_B shown in the above example, perform scalar and matrix multiplication of Matrix_B*Matrix_A and compare the result found in the examples.

5.3.3 Transpose, Inverse, and Determinant of Matrix

Consider the following matrix:

$$A = \begin{bmatrix} 7 & 12 & 5 \\ 14 & 8 & -2 \\ 21 & -15 & 3 \end{bmatrix}.$$

Transpose, inverse, and determinant of the matrix can easily be found out using the default functions in Excel for matrices.

To transpose the matrix, select the cells with the same order and type =TRANSPOSE(Matrix_A). Press "Ctrl+Shift+Enter."

SIN	▾	:	✕	✓	fx	=TRANSPOSE(Matrix_A)

	A	B	C	D	E	F
1	7	12	5			
2	14	8	-2			
3	21	-15	3			
4						
5	=TRANSPOSE(Matrix_A)					
6						
7						
8						

Fig. 5.39

Matrix_B	▾	:	✕	✓	fx	{=TRANSPOSE(Matrix_A)}

	A	B	C	D	E	F
1	7	12	5			
2	14	8	-2			
3	21	-15	3			
4						
5	7	14	21			
6	12	8	-15			
7	5	-2	3			

Fig. 5.40

To inverse the matrix, select the cells with the same order and type =MINSVERSE(Matrix_A). Press "Ctrl+Shift+Enter."

SIN	▾	:	✕	✓	fx	=MINVERSE(Matrix_A)

	A	B	C	D	E	F
1	7	12	5			
2	14	8	-2			
3	21	-15	3			
4						
5	=MINVERSE(Matrix_A)					
6						
7						
8						

Fig. 5.41

Matrix_B	▾	:	×	✓	*fx*	{=MINVERSE(Matrix_A)}

	A	B	C	D	E	F
1	7	12	5			
2	14	8	-2			
3	21	-15	3			
4						
5	0.002041	0.037755	0.021769			
6	0.028571	0.028571	-0.02857			
7	0.128571	-0.12143	0.038095			
8						

Fig. 5.42

To find out the determinant of the matrix, select a cell (as the determinant is a value, not a matrix) and type =MDETERM(Matrix_A). As the function will return a value, not a matrix, there is no need to press "Ctrl+Shift+Enter." Pressing "Enter" only will execute the function.

SIN	▾	:	×	✓	*fx*	=MDETERM(Matrix_A)

	A	B	C	D	E	F
1	7	12	5			
2	14	8	-2			
3	21	-15	3			
4						
5	=MDETERM(Matrix_A)					
6						

Fig. 5.43

A6	▾	:	×	✓	*fx*	

	A	B	C	D
1	7	12	5	
2	14	8	-2	
3	21	-15	3	
4				
5	-2940			
6				

Fig. 5.44

Exercise 5.4: Perform the transpose and inverse operation on the following matrix. Find the determinant of each matrix.

$$A = \begin{bmatrix} 32 & -10 & 0 \\ -9 & 11 & 4 \\ 3 & -16 & 12 \end{bmatrix}.$$

5.4   Equations

Microsoft Excel is specially designed to deal with equations. From the scattered points, not only Excel can draw the line, but if required, can extrapolate to forecast the value from the existing set. Moreover, various strategies of numerical methods can be implemented in Excel, either by using built-in GUI or by coding in VBA.

5.4.1 Plotting Equations in Excel

If two equations are given as $Y = x^3 - 5x^2 + x - 4$ and $Z = 37x^2 - 12x + 96$, then the graph is to be plotted for $0 < x < 50$. Follow the instructions to plot the equations.

1. Enter Headings in columns X, Y, and Z.

2. In the first column, enter the x values. There is a quick trick to fill the column with numbers from 0 to 50 faster than typing by doing as follows:
   * Enter 1 in A2 and 2 in A3.
   * Highlight A2 and A3.
   * Move the cursor to the bottom left until a plus sign appears.
   * Press, hold, and drag the cursor down until a 30 appears.

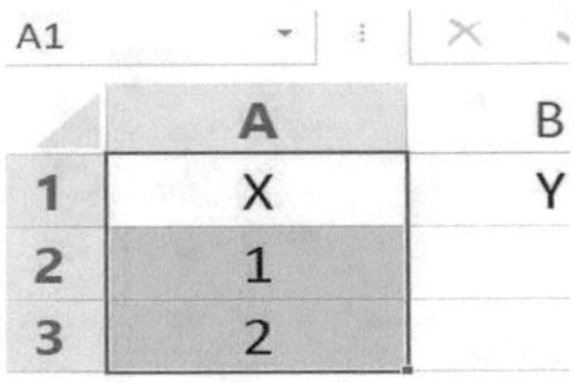

Fig. 5.45

3. In B2, enter the formula +A2^3-5*A2^2+A2-4.
4. In C2, enter the formula +37*A2^2-12*A2+96.

   In the above formula, + at the starting represents a formula. Whenever we want to use a formula, we will start with the + sign. To fill in the rest of the table:

(i) Highlight cells B2 and C2 with the mouse.

(ii) Hover the cursor on small plus sign at the bottom right of the cell C2.
(iii) Then, press and drag the cursor downwards to the last row.

Result:

	A	B	C	D
1	X	Y	Z	
2	1	-7	121	
3	2	-14	220	
4	3	-19	393	
5	4	-16	640	
6	5	1	961	
7	6	38	1356	
8	7	101	1825	
9	8	196	2368	
10	9	329	2985	
11	10	506	3676	
12	11	733	4441	
13	12	1016	5280	
14	13	1361	6193	
15	14	1774	7180	
16	15	2261	8241	
17	16	2828	9376	
18	17	3481	10585	
19	18	4226	11868	
20	19	5069	13225	
21	20	6016	14656	
22	21	7073	16161	
23	22	8246	17740	
24	23	9541	19393	
25	24	10964	21120	
26	25	12521	22921	
27	26	14218	24796	

Sheet1 (+)

READY

Fig. 5.46

Graph the Data

Select A1:C51. Then click "Insert" and select "Scatter with Smooth Lines." The graph is given below.

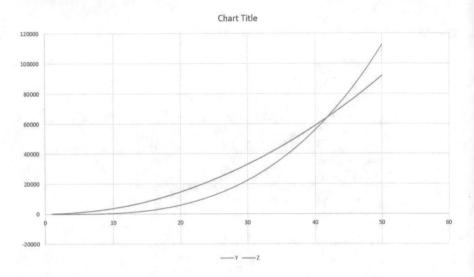

Fig. 5.47

Exercise 5.5: Plot the sin(θ) function from $0 < \theta < 180$ (degrees) at 15 degrees interval.

5.4.2   Trendlines

For several scatter points on a graph, we may need to determine a line or curve which will almost follow the trend of the data so that we can predict the future value by extending the line. That particular line is called the trendline in Excel. This example demonstrates the process of adding a trendline to a chart. For the following data:

X	Y
4	3
2	2
9	5
6	7
1	2
0	3
4	6
2	5
7	6
9	4

1. Copy the data in Excel, select the data and plot the Scatter Chart. It should look as follows:

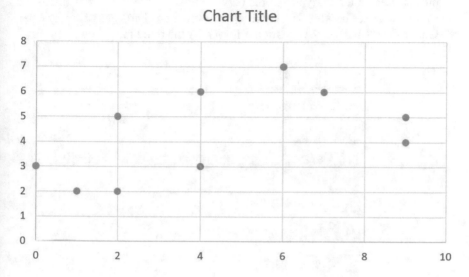

Fig. 5.48

2. Click the + button beside the chart. Then click the arrow beside the trendline. Click More Options.

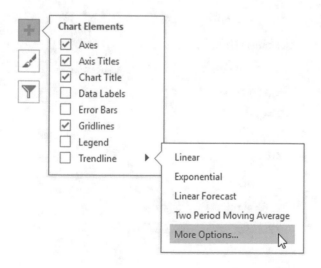

Fig. 5.49

The Format Trendline pane pops up.

3. From Trendline options, click Linear.
4. Affix the number of periods in the Forecast Forward option. For instance, to predict the data for 15, type 5 in the Forward box.
5. Check the boxes beside "Display R-squared value on chart" and "Display equation on chart" and see the equation followed by the trendline.

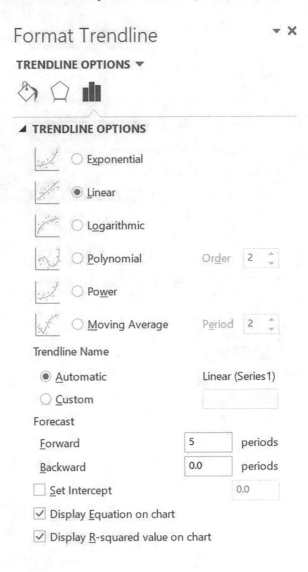

Fig. 5.50

Result:

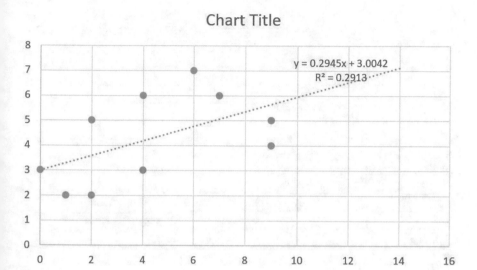

Fig. 5.51

Exercise 5.6: Plot this scatter plot and add a trendline to it. Find the governing/characteristics equation of Sales. Find the number of Sales in 2020 using both linear fitting and polynomial fitting.

Years	Sales
2011	$23,984
2012	$17,348
2013	$14,873
2014	$26,349
2015	$25,384

5.5   Forecasting

Excel allows us to predict missing or future data based on a given set of data. Excel follows a linear trend to forecast data using the FORECAST function. The FORECAST.ETS function can predict an unknown value via the method of exponential triple smoothing, taking seasonality into consideration [1]. The FORECAST function is quite old. The new FORECAST.LINEAR function is a newer and better alternative.

FORECAST.LINEAR

1. The function FORECAST.LINEAR can predict a future value in a linear fashion.

SUM	▾	⋮	✕	✓	*fx*	=FORECAST.LINEAR(A8,B2:B7,A2:A7)

◢	A	B	C	D	E
1	Time (Days)	Demand (MW)	Forecast		
2	1	30			
3	2	35			
4	3	48			
5	4	57			
6	5	72			
7	6	75			
8	7		=FORECAST.LINEAR(A8,B2:B7,A2:A7)		
9	8		FORECAST.LINEAR(x, known_ys, known_xs)		
10	9		106.2		
11	10		113		
12					

Fig. 5.52

By dragging the FORECAST.LINEAR function downwards, the absolute references ($B$2:$B$7 and $A$2:$A$7) remain fixed. However, the relative reference cell (A8) changes to A9 and A10.

2. Enter 75 into cell C7. Select the cells A1:C11. Create a scatter plot with straight
   lines and markers.

Chart 3	▾	⋮	✕	✓	*fx*	

◢	A	B	C
1	Time (Days)	Demand (MW)	Forecast
2	1	30	
3	2	35	
4	3	48	
5	4	57	
6	5	72	
7	6	75	75
8	7		87.33333333
9	8		99
10	9		106.2
11	10		113
12			

Fig. 5.53

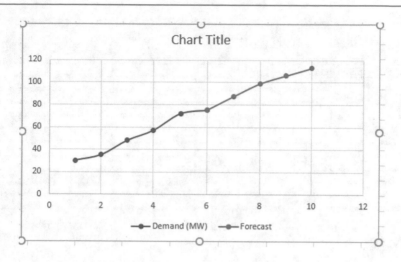

Fig. 5.54

When a trendline is added to a chart, the equation can be displayed. The equation can extrapolate future values based on the given data.

FORECAST.ETS

The FORECAST.ETS function is available in Excel 2016 or later versions. This function can detect a seasonal pattern of data.

1. The function predicts a future value by Exponential Triple Smoothing.

SUM	▼	⋮	×	✓	fx	=FORECAST.ETS(A8,B2:B7,A2:A7,1)

◢	A	B	C	D	E	F	G	H	I	J
1	Time (Days)	Demand (MV	Forecast							
2	1	45								
3	2	35								
4	3	48								
5	4	32								
6	5	58								
7	6	42								
8	7		=FORECAST.ETS(A8,B2:B7,A2:A7,1)							
9	8		FORECAST.ETS(target_date, values, timeline, [seasonality], [data_completion], [aggregation])							
10	9		50.90149							
11	10		72.42222							

Fig. 5.55

Providing the last three arguments is not mandatory. The fourth argument defines the size of the seasonal pattern. A default value of 1 implies that the seasonality is automatically found.

2. Enter 42 in the cell C. Select the cells A1:C11. Create a scatter plot with straight lines and markers.

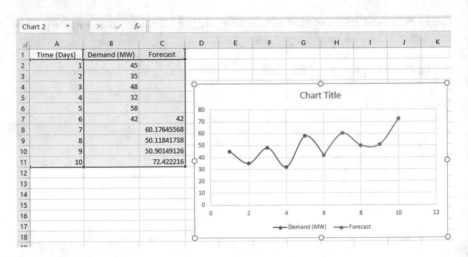

Fig. 5.56

3. The function FORECAST.ETS.SEASONALITY can be used to find the length of the seasonal pattern.

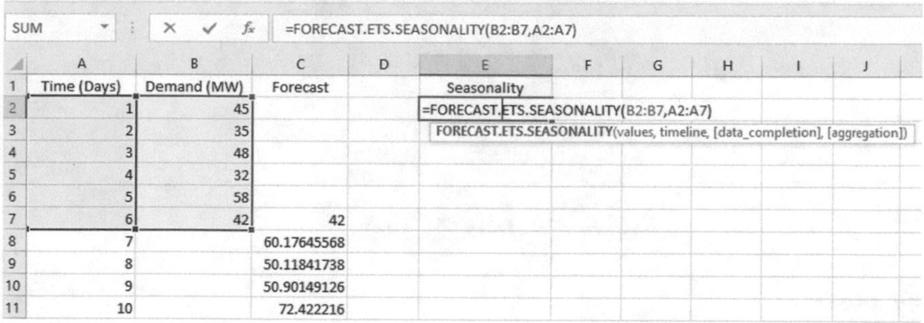

Fig. 5.57

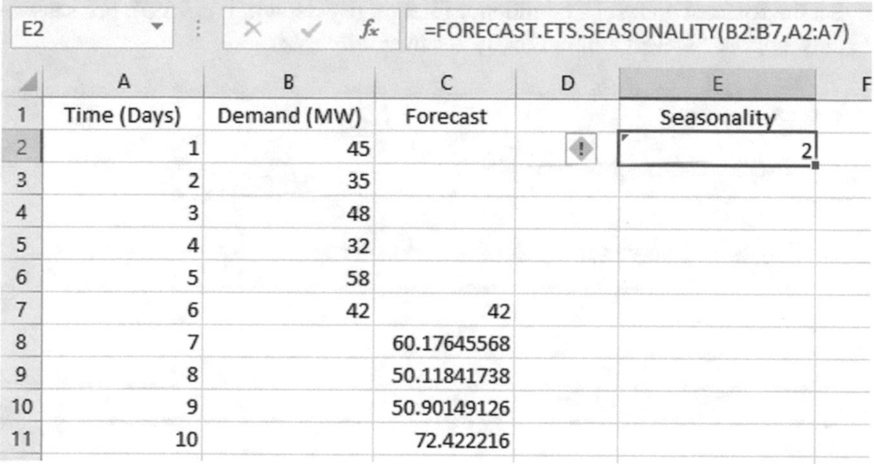

Fig. 5.58

Forecast Sheet

Forecast Sheet can be found in Excel 2016 and later versions to automatically create a visual forecast worksheet.

1. Select the range of cells containing the target data.

2. From the ribbon, go to Data > Forecast > Forecast Sheet.

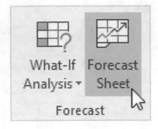

Fig. 5.59

The dialog box shown in Fig. 5.60 appears.

3. Set the forecast end; set a confidence interval (by default, it is 95%). For season-ality, choose "Detect automatically" or manually set it.

Fig. 5.60

4. Click on Create.

The Forecast Sheet tool uses the FORECAST.ETS function to predict the future values. The upper and lower confidence bounds are an added feature.

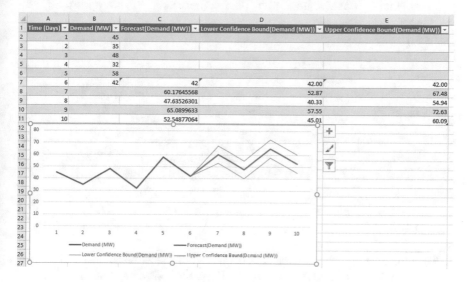

Fig. 5.61

5.6   Data Manipulation

5.6.1   Data Sorting

MS Excel allows alphabetical or numerical data sorting. Data can be sorted in an entire worksheet or in a specific range of cells containing data. *Sort Sheet* can sort the data in a single column of the entire worksheet. The corresponding rows are also shifted accordingly when Sort Sheet is applied. Sort Range can sort the data in only a selected range of cells and leave all other cells unaffected.

Sorting an Entire Worksheet:
In the following example, a Shirt order form will be sorted alphabetically by *the Last Name*.

1. Select a cell from the column by which the data will be sorted. Here, the cell B2 is selected.

◢	A	B	C	D	E
1	**First Name**	**Last Name**	**Shirt Size**	**Quantity**	**Payment Method**
2	Malik	Reynolds	Small	8	Cash
3	Karla	Nichols	X-Large	6	Money Order
4	Singh	Sachdeva	Small	6	Check
5	Pal	Sharma	X-Large	5	Cash
6	Juan	Flores	X-Large	4	Cash
7	Kumar	Mehta	Small	4	Cash
8	Kumar	Gupta	X-Large	4	Debit Card
9	Melissa	White	Small	3	Debit Card
10	Esther	Yaron	Small	2	Cash
11					

Fig. 5.62

2. From the data tab, click on the A–Z command or the Z–A command in order to sort from A to Z or Z to A, respectively. Here, the A–Z command is chosen.

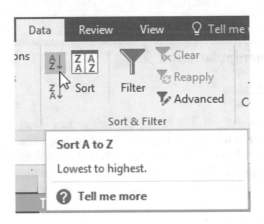

Fig. 5.63

3. The worksheet thus gets sorted by the chosen column B or the last names.

⊿	A	B	C	D	E
1	**First Name**	**Last Name**	**Shirt Size**	**Quantity**	**Payment Method**
2	Juan	Flores	X-Large	4	Cash
3	Kumar	Gupta	X-Large	4	Debit Card
4	Kumar	Mehta	Small	4	Cash
5	Karla	Nichols	X-Large	6	Money Order
6	Malik	Reynolds	Small	8	Cash
7	Singh	Sachdeva	Small	6	Check
8	Pal	Sharma	X-Large	5	Cash
9	Melissa	White	Small	3	Debit Card
10	Esther	Yaron	Small	2	Cash
11					

Fig. 5.64

Sorting a Particular Data Range:

1. Select the desired cell range to be sorted. This example selects the range *A1:E10*.

⊿	A	B	C	D	E
1	**First Name**	**Last Name**	**Shirt Size**	**Quantity**	**Payment Method**
2	Juan	Flores	X-Large	4	Cash
3	Kumar	Gupta	X-Large	4	Debit Card
4	Kumar	Mehta	Small	4	Cash
5	Karla	Nichols	X-Large	6	Money Order
6	Malik	Reynolds	Small	8	Cash
7	Singh	Sachdeva	Small	6	Check
8	Pal	Sharma	X-Large	5	Cash
9	Melissa	White	Small	3	Debit Card
10	Esther	Yaron	Small	2	Cash
11					

Fig. 5.65

2. Go to Data > Sort.

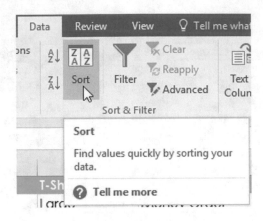

Fig. 5.66

3. From the dialog box named Sort, select a column to sort by. In this example, the number of shirt orders is the sorting criterion, hence the option "Quantity" is chosen.

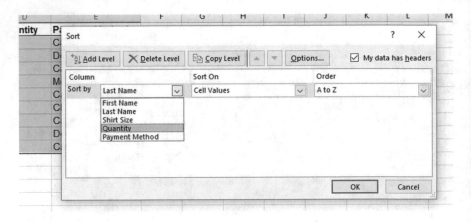

Fig. 5.67

4. Choose the order for sorting (descending or ascending). Here, the order is Largest to Smallest. Click OK.

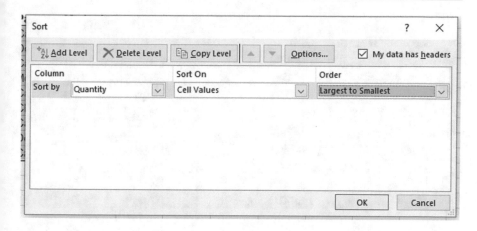

Fig. 5.68

Result: The cells are sorted by the column of choice. Here, the column Quantity will
be sorted from the highest to the lowest quantity of order.

	A	B	C	D	E
1	**First Name**	**Last Name**	**Shirt Size**	**Quantity**	**Payment Method**
2	Malik	Reynolds	Small	8	Cash
3	Karla	Nichols	X-Large	6	Money Order
4	Singh	Sachdeva	Small	6	Check
5	Pal	Sharma	X-Large	5	Cash
6	Juan	Flores	X-Large	4	Cash
7	Kumar	Gupta	X-Large	4	Debit Card
8	Kumar	Mehta	Small	4	Cash
9	Melissa	White	Small	3	Debit Card
10	Esther	Yaron	Small	2	Cash

Fig. 5.69

Exercise 5.7: Sort this list according to the Grade in Column C.

▲	A	B	C	D	E
1	**Camper Name**	**Gender**	**Grade**	**Cabin Color**	**Counselor**
2	Kacey Cranston	Female	6	Pink	Stone
3	John Gibbs	Male	8	Orange	Smith
4	Susana Jimenez	Female	8	Yellow	Chang
5	Flora Jones	Female	7	Green	Gorecki
6	Tia Carter	Female	6	Pink	Stone
7	Miles Goldstein	Male	8	Orange	Smith
8	Taquan Holder	Male	7	Black	Patel
9	Mariela Flores	Female	9	Red	Leslie
10	Priya Dwivedi	Female	9	Red	Leslie

Fig. 5.70

5.6.2   Data Filtering

Too much content can make a worksheet clumsy and hard to navigate. In order to narrow down the data in a worksheet, filters can be used in Excel, which allows us to navigate quickly to the desired information. In the example that follows, a filter is applied to a worksheet for showing only the information of the sold items.

1. The worksheet should contain a header row that identifies each column.

▲	A	B	C	D	E
1	**Invoice ID**	**Customer Name**	**Equipment**	**Equipment Detail**	**Sold Date**
2	3000	Shannon Nguyen	Camera	Saris Lumina Digital Camera	12-May-19
3	3005	Sela Shepard	Camera	Saris Zoom Z-60 Digital Camera	27-Jul-19
4	1031	Nick Ortiz	Laptop	17" Saris X-10 Laptop	04-Oct-19
5	1021	Sofie Ragnar	Laptop	15" EDI SmartPad L200-3 Laptop	15-Sep-19
6	5022	Carl Langer	TV	32" Paragon 440 OLED TV	17-Jul-19
7	5023	Margaret Lisbon	TV	50" Paragon 490L LED TV	01-Oct-19
8	6102	Jamila Kyle	Projector	Omega VisX 1.0	22-Aug-19
9	6200	Jolie Chaturvedi	Projector	Saris Lux T-80	01-Sep-19
10	1012	August Zorn	Tablet	Saris SlimPro	29-Sep-19
11	2051	Sofie Ragnar	Other	EDI SmartBoard L500-1	01-Oct-19
12	3800	Hank Sorenson	Other	U-Go Saris DigiCam Printer II	04-Aug-19
13	3900	Clint Gosse	Other	U-Go Saris Label Maker	13-Jun-19
14	4800	Sela Shepard	Other	7N Deluxe Camera Travel Bag	27-Jul-19
15					

Fig. 5.71

2. From the Data tab, select the Filter command. Then, a drop-down menu will be available in the label or header cell of each column. This step is common for all filtering techniques in Excel. Remember this as the *initial filtering step*.

Fig. 5.72

3. Expand the drop-down menu of the column which is to be filtered. Here, column
   C is to be filtered to choose specific types of equipment.

	A	B	C	D
1	Invoice ID ▾	Customer Name ▾	Equipment ▾	Equipment I
2	3000	Shannon Nguyen	Camera	Saris Lumina Digit
3	3005	Sela Shepard	Camera	om Z-60 Dig
4	1031	Nick Ortiz	Laptop	17" Saris X-10
5	1021	Sofie Ragnar	Laptop	15" EDI SmartPad L2

Equipment:
(Showing All)

Fig. 5.73

4. A Filter menu pops up.

5. Deselect all data at once by unchecking the box for Select All.

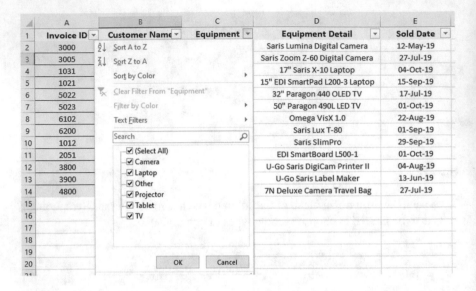

Fig. 5.74

6. Check the boxes beside the data to be filtered out. Click OK. Here, the types
   Laptop and Projector are checked for viewing.

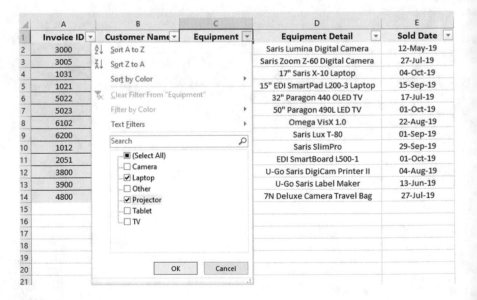

Fig. 5.75

7. The data is filtered. All equipments other than laptop and projector are hidden
   from the list.

	A	B	C	D	E
1	Invoice ID ▾	Customer Name ▾	Equipment ⌄ᴛ	Equipment Detail ▾	Sold Date ▾
4	1031	Nick Ortiz	Laptop	17" Saris X-10 Laptop	04-Oct-19
5	1021	Sofie Ragnar	Laptop	15" EDI SmartPad L200-3 Laptop	15-Sep-19
8	6102	Jamila Kyle	Projector	Omega VisX 1.0	22-Aug-19
9	6200	Jolie Chaturvedi	Projector	Saris Lux T-80	01-Sep-19
15					

Fig. 5.76

The filtering options are also available from Home > Sort & Filter.

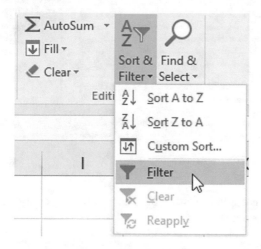

Fig. 5.77

Applying Multiple Filters:

Multiple filters can be applied in Excel to narrow down the results and filter out data. Such Excel filters are called *cumulative filters*. In the previous example, we've already filtered laptops and projectors. Here, the data will be further narrowed down to only show the laptops and projectors sold in the month of August.

1. After the initial filtering step, click on the drop-down menu for the new column for filtering. Here, we shall filter column E to sort the data by date of selling.

	A	B	C	D	E	F
1	Invoice ID ▾	Customer Name ▾	Equipment ⌄ᴛ	Equipment Detail ▾	Sold Date ▾	
4	1031	Nick Ortiz	Laptop	17" Saris X-10 Laptop	04-Oct-19	Sold Date:
5	1021	Sofie Ragnar	Laptop	15" EDI SmartPad L200-3 Laptop	15-Sep-19	(Showing All)
8	6102	Jamila Kyle	Projector	Omega VisX 1.0	22-Aug-19	
9	6200	Jolie Chaturvedi	Projector	Saris Lux T-80	01-Sep-19	

Fig. 5.78

2. The Filter menu appears. Depending on the data to be filtered, check or uncheck the boxes. Click OK. Here, all available months are unchecked except August.

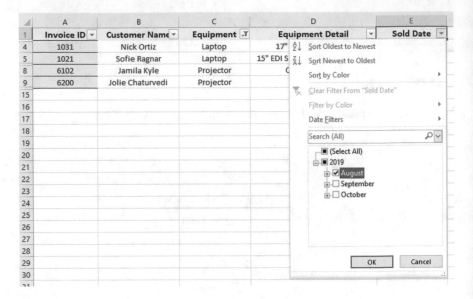

Fig. 5.79

3. The new filter is applied. Thus, the worksheet displays only the laptops and projectors sold in August.

	A	B	C	D	E
1	Invoice ID ▾	Customer Name ▾	Equipment ⅂ ▾	Equipment Detail ▾	Sold Date ⅂
8	6102	Jamila Kyle	Projector	Omega VisX 1.0	22-Aug-19
15					

Fig. 5.80

Clearing a Filter

Go to Data > Sort & Filter > Filter for removing the filters from the worksheet.

Fig. 5.81

Using Advanced Number Filters

Data can be manipulated in different ways using Advanced number filters. The following example shows the process of displaying only certain equipment types based on a given range of Invoice IDs.

1. After the initial filtering step, expand the drop-down menu in the column to be filtered. In this example, column A contains the Invoice ID, which is the basis for filtering this dataset.

	A	B	C	D	E
1	**Invoice ID** ▼	**Customer Name** ▼	**Equipment** ▼	**Equipment Detail** ▼	**Sold Date** ▼
2	3000	Invoice ID: on Nguyen	Camera	Saris Lumina Digital Camera	12-May-19
3	3900	(Showing All) t Gosse	Other	U-Go Saris Label Maker	13-Jun-19
4	5022	Carl Langer	TV	32" Paragon 440 OLED TV	17-Jul-19
5	3005	Sela Shepard	Camera	Saris Zoom Z-60 Digital Camera	27-Jul-19
6	4800	Sela Shepard	Other	7N Deluxe Camera Travel Bag	27-Jul-19
7	3800	Hank Sorenson	Other	U-Go Saris DigiCam Printer II	04-Aug-19
8	6102	Jamila Kyle	Projector	Omega VisX 1.0	22-Aug-19
9	6200	Jolie Chaturvedi	Projector	Saris Lux T-80	01-Sep-19
10	1021	Sofie Ragnar	Laptop	15" EDI SmartPad L200-3 Laptop	15-Sep-19
11	1012	August Zorn	Tablet	Saris SlimPro	29-Sep-19
12	5023	Margaret Lisbon	TV	50" Paragon 490L LED TV	01-Oct-19
13	2051	Sofie Ragnar	Other	EDI SmartBoard L500-1	01-Oct-19
14	1031	Nick Ortiz	Laptop	17" Saris X-10 Laptop	04-Oct-19

Fig. 5.82

2. The Filter menu pops up. Go to Number Filters and select the desired number filter. This example selects Between… to display Invoice IDs within a particular range.

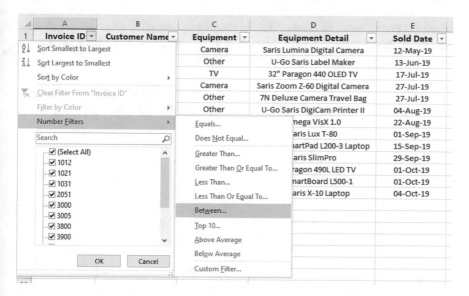

Fig. 5.83

3. The dialog box Custom AutoFilter appears. Insert the desired range of Invoice IDs, and click OK.

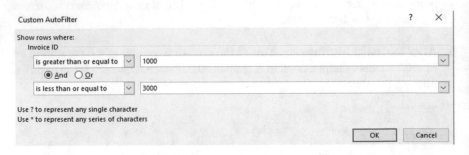

Fig. 5.84

4. The data is thus filtered by the chosen number filter. Here, only the Invoice IDs between 1000 and 3000 are displayed.

	A	B	C	D	E
1	Invoice ID ⟋	Customer Name ▾	Equipment ▾	Equipment Detail ▾	Sold Date ▾
2	1012	August Zorn	Tablet	Saris SlimPro	29-Sep-19
10	1021	Sofie Ragnar	Laptop	15" EDI SmartPad L200-3 Laptop	15-Sep-19
11	1031	Nick Ortiz	Laptop	17" Saris X-10 Laptop	04-Oct-19
13	2051	Sofie Ragnar	Other	EDI SmartBoard L500-1	01-Oct-19
14	3000	Shannon Nguyen	Camera	Saris Lumina Digital Camera	12-May-19

Fig. 5.85

Using Advanced Date Filters

Advanced date filters can display data belonging to a specific time period, such as last month, next year, or between two dates. Here, advanced date filters are used to show only the equipment sold from August 1 to October 1.

1. After the initial filtering step, click on the drop-down menu of the desired column. Here, column E will be filtered to display only a specific range of dates.

⊿	A	B	C	D	E
1	**Invoice ID** ▾	**Customer Name** ▾	**Equipment** ▾	**Equipment Detail** ▾	**Sold Date** ▾
2	3000	Shannon Nguyen	Camera	Saris Lumina Digital Camera	12-May-19
3	3005	Sela Shepard	Camera	Saris Zoom Z-60 Digital Camera	27-Jul-19
4	1031	Nick Ortiz	Laptop	17" Saris X-10 Laptop	04-Oct-19
5	1021	Sofie Ragnar	Laptop	15" EDI SmartPad L200-3 Laptop	15-Sep-19
6	5022	Carl Langer	TV	32" Paragon 440 OLED TV	17-Jul-19
7	5023	Margaret Lisbon	TV	50" Paragon 490L LED TV	01-Oct-19
8	6102	Jamila Kyle	Projector	Omega VisX 1.0	22-Aug-19
9	6200	Jolie Chaturvedi	Projector	Saris Lux T-80	01-Sep-19
10	1012	August Zorn	Tablet	Saris SlimPro	29-Sep-19
11	2051	Sofie Ragnar	Other	EDI SmartBoard L500-1	01-Oct-19
12	3800	Hank Sorenson	Other	U-Go Saris DigiCam Printer II	04-Aug-19
13	3900	Clint Gosse	Other	U-Go Saris Label Maker	13-Jun-19
14	4800	Sela Shepard	Other	7N Deluxe Camera Travel Bag	27-Jul-19

Fig. 5.86

2. From the Filter menu, go to Date Filters, and choose a date filter from the drop-down menu. Here, Between is selected to display equipment sold between August 1 and October 1.

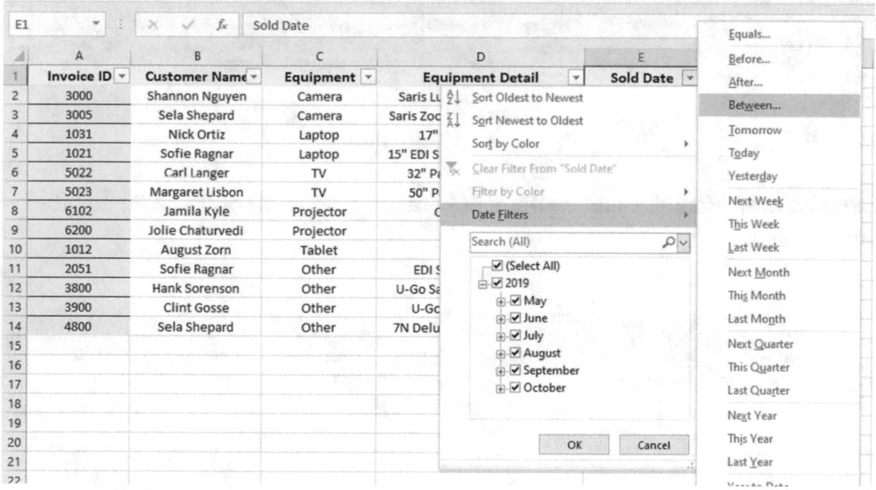

Fig. 5.87

3. In the Custom AutoFilter dialog box, insert desired dates beside each filter and click OK. Here, it is desired to filter the dates outside August 1, 2019 to October 1, 2019.

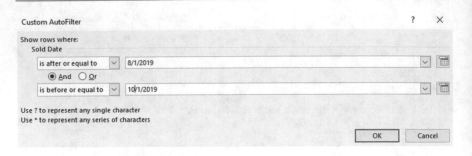

Fig. 5.88

4. The data in the worksheet is thus filtered by the chosen date filter. Here, the items checked out between August 1 and October 1 are now displayed.

	A	B	C	D	E
1	**Invoice ID**	**Customer Name**	**Equipment**	**Equipment Detail**	**Sold Date**
7	3800	Hank Sorenson	Other	U-Go Saris DigiCam Printer II	04-Aug-19
8	6102	Jamila Kyle	Projector	Omega VisX 1.0	22-Aug-19
9	6200	Jolie Chaturvedi	Projector	Saris Lux T-80	01-Sep-19
10	1021	Sofie Ragnar	Laptop	15" EDI SmartPad L200-3 Laptop	15-Sep-19
11	1012	August Zorn	Tablet	Saris SlimPro	29-Sep-19
12	5023	Margaret Lisbon	TV	50" Paragon 490L LED TV	01-Oct-19
13	2051	Sofie Ragnar	Other	EDI SmartBoard L500-1	01-Oct-19

Fig. 5.89

Exercise 5.8: Find the sold equipment between Invoice ID 5000–6000 & July 1 – September 1.

	A	B	C	D	E
1	**Invoice ID**	**Customer Name**	**Equipment**	**Equipment Detail**	**Sold Date**
2	3000	Shannon Nguyen	Camera	Saris Lumina Digital Camera	12-May-19
3	3005	Sela Shepard	Camera	Saris Zoom Z-60 Digital Camera	27-Jul-19
4	1031	Nick Ortiz	Laptop	17" Saris X-10 Laptop	04-Oct-19
5	1021	Sofie Ragnar	Laptop	15" EDI SmartPad L200-3 Laptop	15-Sep-19
6	5022	Carl Langer	TV	32" Paragon 440 OLED TV	17-Jul-19
7	5023	Margaret Lisbon	TV	50" Paragon 490L LED TV	01-Oct-19
8	6102	Jamila Kyle	Projector	Omega VisX 1.0	22-Aug-19
9	6200	Jolie Chaturvedi	Projector	Saris Lux T-80	01-Sep-19
10	1012	August Zorn	Tablet	Saris SlimPro	29-Sep-19
11	2051	Sofie Ragnar	Other	EDI SmartBoard L500-1	01-Oct-19
12	3800	Hank Sorenson	Other	U-Go Saris DigiCam Printer II	04-Aug-19
13	3900	Clint Gosse	Other	U-Go Saris Label Maker	13-Jun-19
14	4800	Sela Shepard	Other	7N Deluxe Camera Travel Bag	27-Jul-19

Fig. 5.90

5.6.3   Flash Fill

In Excel 2013 or later versions, Flash Fill is used to automatically extract or combine data. This feature only works when Excel recognizes a pattern. For instance, the flash fill can be used to combine the last names given in column A with the first names given in column B to generate an email address for each person.

1. First, enter the correct email address in cell C1.

	A	B	C
1	**First Name**	**Last Name**	**Email Adress**
2	Malik	Reynolds	reynolds.malik@organization.com
3	Karla	Nichols	
4	Singh	Sachdeva	
5	Pal	Sharma	
6	Juan	Flores	
7	Kumar	Gupta	
8	Melissa	White	
9	Esther	Yaron	

Fig. 5.91

2. From Data > Data Tools, click on Flash Fill.

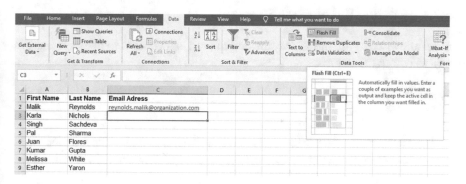

Fig. 5.92

Result:

▲	A	B	C
1	**First Name**	**Last Name**	**Email Adress**
2	Malik	Reynolds	reynolds.malik@organization.com
3	Karla	Nichols	nichols.karla@organization.com
4	Singh	Sachdeva	sachdeva.singh@organization.com
5	Pal	Sharma	sharma.pal@organization.com
6	Juan	Flores	flores.juan@organization.com
7	Kumar	Gupta	gupta.kumar@organization.com
8	Melissa	White	white.melissa@organization.com
9	Esther	Yaron	yaron.esther@organization.com
10			

Fig. 5.93

5.7   Data Cleaning Techniques

Data cleaning in MS Excel includes many variants, such as removal of extra or blank spaces and duplicate data, changing text to upper or lower case, converting stored text into numbers, highlighting errors, spell check, parsing data, etc. [2]. Five amazing data cleaning techniques are described below.

5.7.1   Getting Rid of Extra Spaces

Excel allows us to use the TRIM function to omit extra spaces in between texts, as well as before or after texts.

Syntax: =TRIM(text)

The TRIM function takes in a cell reference (or text) as input. The function omits any spaces before, after, or in between the text (except single spaces between words).

▲	A	B
1	Trim Function will remove the extra spaces	=TRIM(A1)
2		

Fig. 5.94

Result:

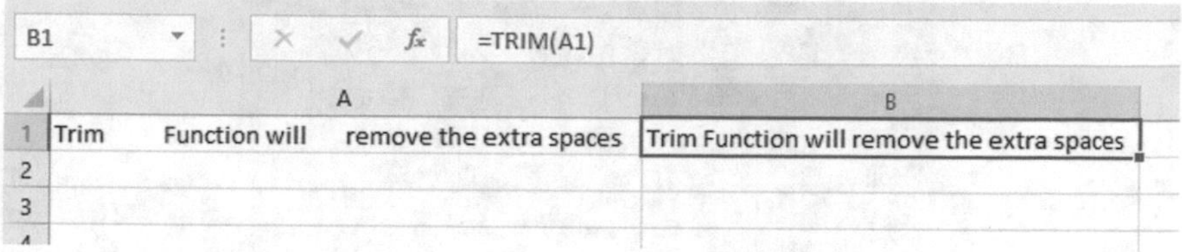

| B1 | ▼ | : | × | ✓ | f_x | =TRIM(A1) |

◢	A	B
1	Trim Function will remove the extra spaces	Trim Function will remove the extra spaces
2		
3		

Fig. 5.95

5.7.2 Handling Blank Cells

Blank cells might occur among a large set of data and can cause a lot of trouble to the user. Finding blank cells is a tedious job if done manually. But Excel has a great way of selecting all blank cells in a given data set simultaneously. The blank cells can be simply highlighted or marked with a certain symbol. The following example shows the method of selecting all blank cells at once.

◢	A	B	C	D
1	Student Name	Math	Chemistry	Physics
2	Bill	87	84	93
3	Mike		80	87
4	Martha	84	78	81
5	Jonas	90		91
6	Magnus	68	78	80
7	Adam		86	72
8	Claudia	92	87	90
9				

Fig. 5.96

1. Select the complete data set.
2. Go to Home > Editing > Find & Select > Go To Special.

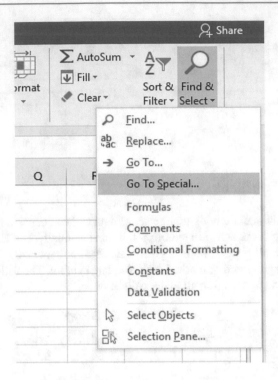

Fig. 5.97

3. Select Blanks. Then click OK.

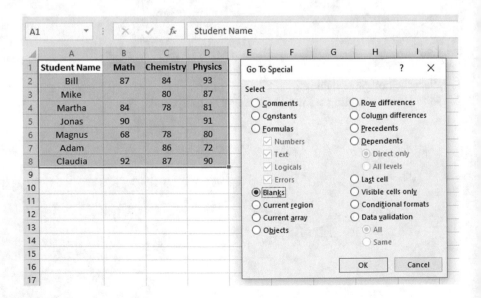

Fig. 5.98

In this way, all blank cells in the data set are selected. Inputs such as 0, Absent, or Not Available can be inserted in such blank cells, only by typing the desired text and pressing CTRL+Enter.

◢	A	B	C	D
1	**Student Name**	**Math**	**Chemistry**	**Physics**
2	Bill	87	84	93
3	Mike		80	87
4	Martha	84	78	81
5	Jonas	90		91
6	Magnus	68	78	80
7	Adam		86	72
8	Claudia	92	87	90
9				
1ᴏ				

Fig. 5.99

Result:

B3		▼	⋮	×	✓	*fx*	Absent

◢	A	B	C	D
1	**Student Name**	**Math**	**Chemistry**	**Physics**
2	Bill	87	84	93
3	Mike	Absent	80	87
4	Martha	84	78	81
5	Jonas	90	Absent	91
6	Magnus	68	78	80
7	Adam	Absent	86	72
8	Claudia	92	87	90
9				

Fig. 5.100

5.7.3 Duplicate Detection and Removal

Finding Duplicates
It is possible to find duplicate data in Excel and to highlight them. The following example demonstrates the process of finding duplicate data in Excel.

1. Select the range of the data in which duplicates are to be searched for. In this case, cells A1:B13 have been selected.

	A	B
1	**First Name**	**Last Name**
2	Malik	Reynolds
3	Karla	Nichols
4	Singh	Sachdeva
5	Pal	Sharma
6	Juan	Flores
7	Kumar	Gupta
8	Kumar	Mehta
9	Melissa	White
10	Esther	Yaron
11	Karla	Nichols
12	Pal	Sharma
13	Kumar	Mehta

Fig. 5.101

2. Go to Home > Styles > Conditional Formatting.
3. Choose option Highlight Cells Rules. Then click on Duplicate Values.

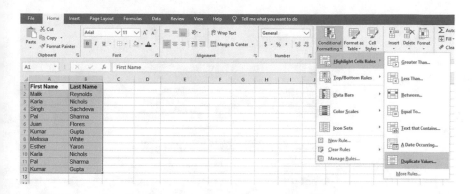

Fig. 5.102

4. Select a formatting style. Click OK.

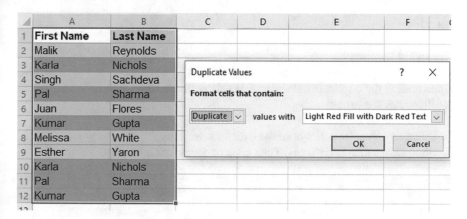

Fig. 5.103

Result: The duplicate names are highlighted.

	A	B
1	**First Name**	**Last Name**
2	Malik	Reynolds
3	Karla	Nichols
4	Singh	Sachdeva
5	Pal	Sharma
6	Juan	Flores
7	Kumar	Gupta
8	Melissa	White
9	Esther	Yaron
10	Karla	Nichols
11	Pal	Sharma
12	Kumar	Gupta

Fig. 5.104

Removing Duplicates

Upon finding the duplicate data, Excel can also remove them. The following example illustrates the process.

1. Select any specific cell within the data set selection.
2. Go to Data tab > Data Tools. Then click on Remove Duplicates.

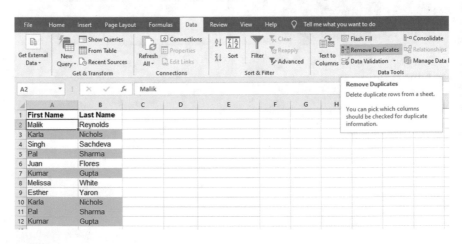

Fig. 5.105

The dialog box in Fig. 5.106 appears.
3. Check on all the checkboxes and then click OK.

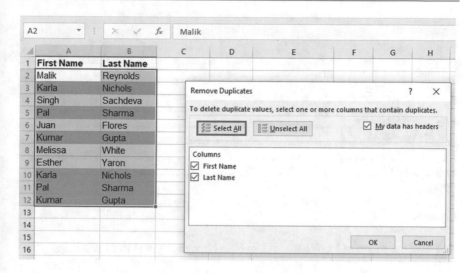

Fig. 5.106

Result: Excel omits all duplicate rows whilst keeping the first one intact.

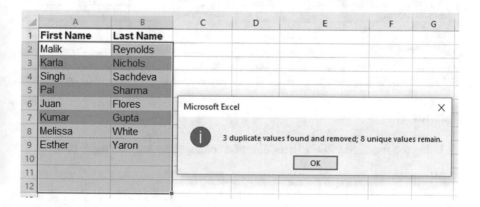

Fig. 5.107

5.7.4 Change Text to Upper, Lower, or Proper Case

The function UPPER(text) can change any text to Upper case.

SUM	▼	⋮	×	✓	fx	=UPPER(A2	

◢	A	B	C	D
1	Name	Upper	Lower	Proper
2	Malik reYnoLds	=UPPER(A2		
3	karla nichols	UPPER(text)		
4	Singh SachDEVA			
5	PAUL SHARMA			
6	JuAN flores			
7	KuMar gupta			
8				

Fig. 5.108

Using the function LOWER(text), we can change any text to Lower case.

SUM	▼	⋮	×	✓	fx	=LOWER(A2	

◢	A	B	C	D
1	Name	Upper	Lower	Proper
2	Malik reYnoLds	MALIK REYNOLDS	=LOWER(A2	
3	karla nichols		LOWER(text)	
4	Singh SachDEVA			
5	PAUL SHARMA			
6	JuAN flores			
7	KuMar gupta			

Fig. 5.109

Using the function PROPER(text), we can change any text to Proper case.

SUM	▼	⋮	×	✓	fx	=PROPER(A2	

◢	A	B	C	D
1	Name	Upper	Lower	Proper
2	Malik reYnoLds	MALIK REYNOLDS	malik reynolds	=PROPER(A2
3	karla nichols			PROPER(text)
4	Singh SachDEVA			
5	PAUL SHARMA			
6	JuAN flores			
7	KuMar gupta			

Fig. 5.110

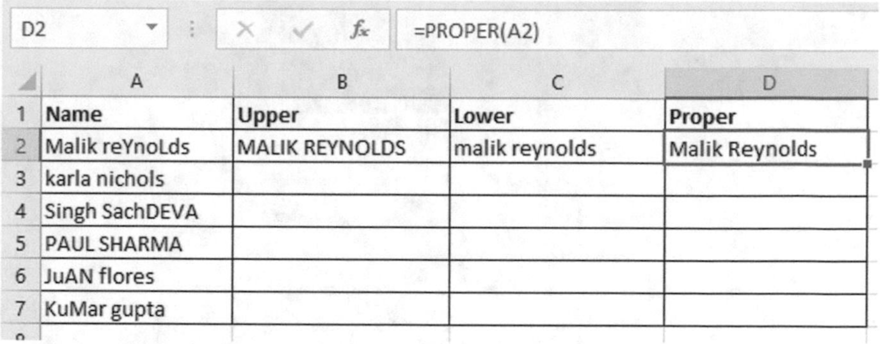

| D2 | ▼ | : | × | ✓ | *fx* | =PROPER(A2) |

◢	A	B	C	D
1	Name	Upper	Lower	Proper
2	Malik reYnoLds	MALIK REYNOLDS	malik reynolds	Malik Reynolds
3	karla nichols			
4	Singh SachDEVA			
5	PAUL SHARMA			
6	JuAN flores			
7	KuMar gupta			

Fig. 5.111

Result:

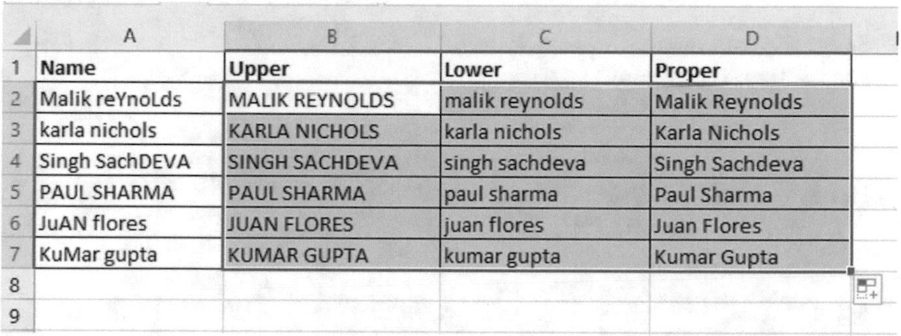

◢	A	B	C	D
1	Name	Upper	Lower	Proper
2	Malik reYnoLds	MALIK REYNOLDS	malik reynolds	Malik Reynolds
3	karla nichols	KARLA NICHOLS	karla nichols	Karla Nichols
4	Singh SachDEVA	SINGH SACHDEVA	singh sachdeva	Singh Sachdeva
5	PAUL SHARMA	PAUL SHARMA	paul sharma	Paul Sharma
6	JuAN flores	JUAN FLORES	juan flores	Juan Flores
7	KuMar gupta	KUMAR GUPTA	kumar gupta	Kumar Gupta
8				
9				

Fig. 5.112

5.7.5   Convert Numbers Stored as Text to Numbers

In the list below, a few costs are not numbers, which is why the total cost is not correct. We have to convert them into numbers.

	×	✓	f_x	=SUM(D2:D7)	

B	C	D	E
		List of Cost	
		122	
		49.99	
		78	
		23.45	
		98.23	
		205.00	
	Total	176.23	

Fig. 5.113

1. Select the data range.
2. Data > Text to Columns.

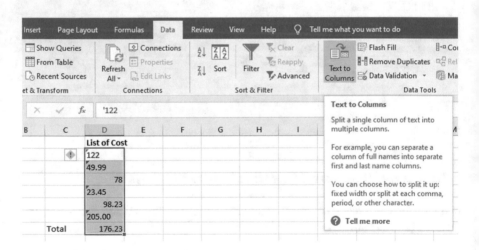

Fig. 5.114

3. Click Finish.

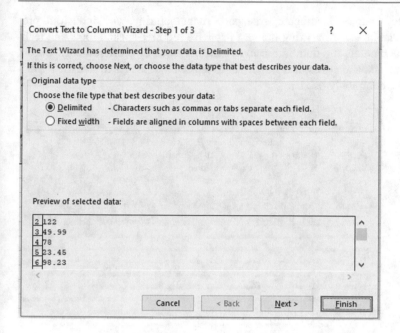

Fig. 5.115

Result:

	C	D
		List of Cost
		122
		49.99
		78
		23.45
		98.23
		205
	Total	576.67

Fig. 5.116

5.8 Application of Excel in Data Analysis

Data science is an emerging interdisciplinary field that unifies computer science, machine learning, and statistics in order to analyze a structured or an unstructured dataset. Data is an essential aspect in all fields of work, including banking, e-commerce, finance, healthcare, transport, manufacturing, and industrial processes, academia, etc. Analyzing the vast database requires special tools and techniques to manipulate the data in the most advantageous or preferred way. Before diving into the data analysis, it is important to learn about the basic statistical functions that are commonly used.

	A	B	C	D	E
1	x	Function Name	Function	Syntax	Result
2	3	AVERAGE	Determines the average	AVERAGE(A2:A16)	4
3	5	COUNT	Count the number of cell with numbers	COUNT(A2:A20)	15
4	1	MEDIAN	Return the median	MEDIAN(A2:A16)	3
5	0	MODE	Return the mode	MODE(A2:A16)	7
6	1	MAX	Displays the maximum number	MAX(A2:A16)	9
7	5	MIN	Displays the minimum number	MIN(A2:A16)	0
8	2	STDEV	Determines the standard deviation	STDEV(A2:A16)	3.047247
9	7				
10	3				
11	7				
12	0				
13	7				
14	8				
15	2				
16	9				

Fig. 5.117

Figure 5.117 represents a dataset from the cells A2 to A16. From this dataset, the AVERAGE (arithmetic mean), the COUNT (number of data), the MEDIAN (middle value of the dataset), the MODE (datum which appears the maximum number of times in the dataset), the MAX (maximum value), the MIN (minimum value), and the STDEV (standard deviation) functions are implemented, and the results are tabulated from cells E2 to E16. All these functions only need the range of the cells in which the data are placed to yield the result. With two rows of data, additional functions can be used to determine the relationship between two sets of data. Figure 5.118 represents the use of the VAR, CORREL, PEARSON, KURT, SKEW, SLOPE, and FORECAST functions upon two columns of data designated by x and y, respectively, representing array1 and array2 in the syntaxes of the functions.

5.8.1   Basic Functions

COVARIANCE
The value of the covariance can be determined by the population covariance or the sample covariance.

$$\text{Population covariance} = \frac{\sum \left(X_i - \bar{X}\right)\left(Y_i - \bar{Y}\right)}{N}.$$

The population covariance can be determined in Excel by using the function COVARIANCE.P(array1, array2).

$$\text{Sample covariance} = \frac{\sum (X_i - \bar{X})(Y_i - \bar{Y})}{N-1}.$$

The sample covariance can be determined in Excel by using the function COVARIANCE.S(array1, array2).

	A	B	C	D	E	F
1	x	y	VAR	Calculates the variation	VAR(A2:A11, B2:B11)	619.5026
2	34	47	CORREL	Determines correlation between two arrays	CORREL(A2:A11,B2:B11)	0.314723
3	65	32	PEARSON	Displays the Pearson Correlation	PEARSON(A2:A11, B2:B11)	0.314723
4	76	80	KURT	Determines the kurtosis	KURT(A2:A11, B2:B11)	-1.03472
5	32	15	SKEW	Determines the skew	SKEW(A2:A11, B2:B11)	-0.41675
6	65	65	SLOPE	Calculates the slope	SLOPE(B2:B11,A2:A11)	0.368565
7	87	89	FORECAST	Forecast the value of y for a given x	FORECAST(100, A2:A11, B2:B11)	66.30357
8	12	88				
9	54	29				
10	76	87				
11	54	66				

Fig. 5.118

VAR

The value of the variance can be determined by the population variance or the sample variance.

$$\text{Population variance, } \sigma^2 = \frac{\sum_{i=1}^{N}(x_i - \mu)^2}{N},$$

where:

σ^2 = population variance;
x_i = value of ith element;
μ = population mean;
N = population size.

The population variance can be determined in Excel by using the function VAR.P(array1, array2).

$$\text{Sample variance, } s^2 = \frac{\sum_{i=1}^{N}(x_i - \bar{x})^2}{n-1},$$

where,

s^2 = sample variance;
x_i = value of ith element;
$\bar{x}$ = sample mean;
n = sample size.

The sample variance can be determined in Excel by using the function VAR.S(array1, array2).

CORREL

$$\text{Correlation coefficient, } r_{xy} = \frac{\Sigma(x_i - \bar{x})(y_i - \bar{y})}{\sqrt{\Sigma(x_i - \bar{x})^2 \, \Sigma(y_i - \bar{y})^2}}.$$

The correlation coefficient of two data arrays can be determined in Excel through the CORREL(array1, array2) function.

PEARSON

$$\text{Pearson correlation coefficient, } r = \frac{n(\Sigma xy) - (\Sigma x)(\Sigma y)}{\sqrt{\left[n\Sigma x^2 - (\Sigma x)^2\right]\left[n\Sigma y^2 - (\Sigma y)^2\right]}},$$

The Pearson product–moment correlation coefficient of two data arrays can be determined in Excel through the PEARSON(array1, array2) function.

KURT

$$\text{Kurtosis} = \frac{\sum_{i=1}^{N} \dfrac{(X_i - \bar{X})}{N}}{s^4},$$

$\bar{X}$ = mean;
s= standard deviation;
N= sample size.

The Kurtosis of a dataset can be determined in Excel by using the KURT(number1, number2, number3…) function.

SKEW

$$\text{Skewness} = \frac{\sum_{i=1}^{N}(X_i - \bar{X})^3}{(N-1)s^3},$$

$\bar{X}$ = mean;
s= standard deviation;
N= sample size.

The Skewness of a dataset can be determined in Excel by using the SKEW(number1, number2, number3…) function.

SLOPE

$$m = \frac{y_2 - y_1}{x_2 - x_1}.$$

FORECAST: Refer to Sect. 5.5.

5.8.2 Advanced Functions

Two advanced functions in MS Excel related to the field of data science are concatenation and VLOOKUP.

Concatenation Function
This function is used to combine texts together and format them accordingly. For example, consider the following dataset:

	A	B	C
1	State	Zip Code	Concatenate
2	Kingston	30145	
3	Colorado Springs	80957	
4	Hamilton	81638	
5	Colorado Springs	89231	
6	Maybell	81640	
7	Oregon	97665	
8	Rydal	30171	
9	Palmer Lake	80133	
10	Peylon	80831	

Fig. 5.119

We need to write the state name and zip code together with a space in the middle. For example: "Kingston 30145."

1. Write the formula "CONCATENATE(A2, B2)" in cell C2.

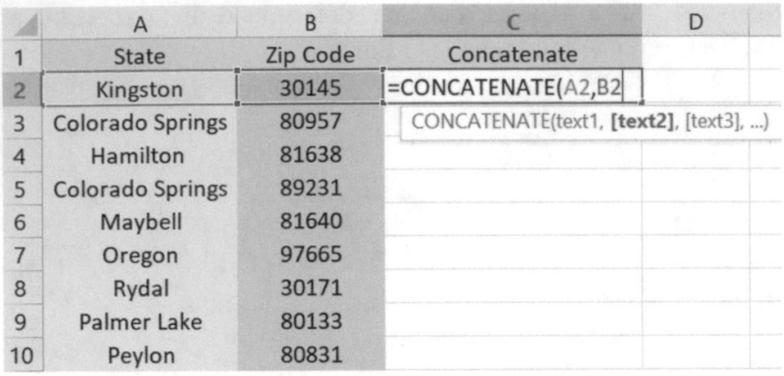

	A	B	C	D
1	State	Zip Code	Concatenate	
2	Kingston	30145	=CONCATENATE(A2,B2	
3	Colorado Springs	80957	CONCATENATE(text1, **[text2]**, [text3], ...)	
4	Hamilton	81638		
5	Colorado Springs	89231		
6	Maybell	81640		
7	Oregon	97665		
8	Rydal	30171		
9	Palmer Lake	80133		
10	Peylon	80831		

Fig. 5.120

2. Pressing enter, the text would appear like this:

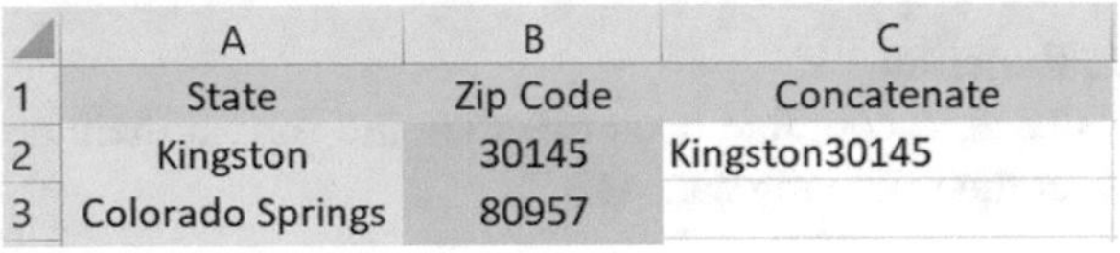

	A	B	C
1	State	Zip Code	Concatenate
2	Kingston	30145	Kingston30145
3	Colorado Springs	80957	

Fig. 5.121

Therefore, we need to add a space in between.

3. Modify the formula in cell C2 as:
   =CONCATENATE(A2," ",A3)

   Where we have added a space string between the quotations.

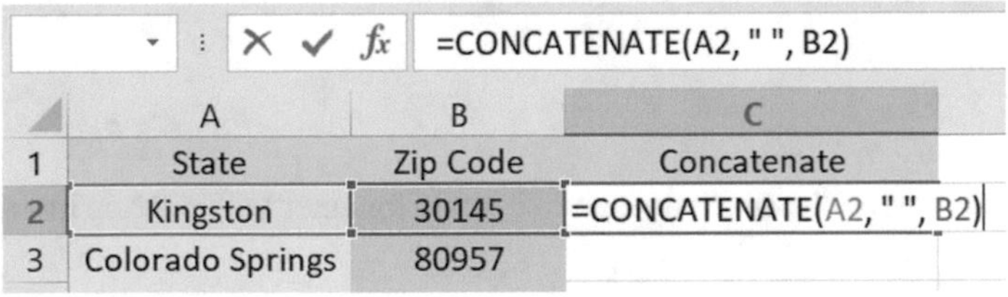

		× ✓ *fx*	=CONCATENATE(A2, " ", B2)	

	A	B	C
1	State	Zip Code	Concatenate
2	Kingston	30145	=CONCATENATE(A2, " ", B2)
3	Colorado Springs	80957	

Fig. 5.122

4. Pressing enter, the right format shows up.

| C2 | ▾ | : | ✕ | ✓ | *fx* | =CONCATENATE(A2, " ", B2) |

◢	A	B	C
1	State	Zip Code	Concatenate
2	Kingston	30145	Kingston 30145
3	Colorado Springs	80957	

Fig. 5.123

5. Lastly, fill flash the format in the following row as well to use the same format for the subsequent cells. The fill flash is performed by bringing the mouse icon in the bold right down corner of the selected box of cell C2 until a "plus" sign appears and drag it to cell C10.

◢	A	B	C
1	State	Zip Code	Concatenate
2	Kingston	30145	Kingston 30145
3	Colorado Springs	80957	Colorado Springs 80957
4	Hamilton	81638	Hamilton 81638
5	Colorado Springs	89231	Colorado Springs 89231
6	Maybell	81640	Maybell 81640
7	Oregon	97665	Oregon 97665
8	Rydal	30171	Rydal 30171
9	Palmer Lake	80133	Palmer Lake 80133
10	Peylon	80831	Peylon 80831
11			

Fig. 5.124

VLOOKUP Function

VLOOKUP, or vertical lookup, is an efficient function for looking up a particular value from a given table array with a keyword of an exact or approximate match. This function is extremely efficient for a long-range of data, but in this example, a small table is shown to comprehend the function syntax better.

Consider the following data:

Player Name	Jersey Number
Jesse	45
Hansen	9
Peter	12
Maxwell	66
Baker	1
Tom	5
Chris	13
Jake	35

Fig. 5.125

Let's say we want to know the Jersey number for Tom (it is obviously 5 from the small table, but we would have difficulty for a large table or multiple sheets).

	A	B	C	D	E	F	G	H	I	J
1	Player Name	Jersey Number								
2	Jesse	45		We want to determine the jersey number of Tom						
3	Hansen	9		Function to be used = VLOOKUP						
4	Peter	12		Result to be put = E9						
5	Maxwell	66								
6	Baker	1								
7	Tom	5								
8	Chris	13								
9	Jake	35		Tom	=VLOOKUP(					
10					VLOOKUP(**lookup_value**, table_array, col_index_num, [range_lookup])					

Fig. 5.126

VLOOKUP function takes four inputs:

(i)   The first argument takes the cell at the value which the function will lookup (in this case, Tom at D8).

(ii) The second argument takes the range of table array (in this case, A2:B9, excluding the titles as we do not need to navigate them).

(iii) The third argument takes the column to be navigated (in this case, column 2 as it has the jersey numbers).

(iv) The last argument takes either 0 or 1. 0 is for FALSE for the exact match for the lookup value (in this case, we want to compare the first column with an exact match with "Tom." 1 would be applicable for an approximate match, for example, if the lookup value had "T," or "tom," or "TOM" instead of the exact keyword).

	A	B	C	D	E	F	G	H	I	J
1	Player Name	Jersey Number								
2	Jesse	45		We want to determine the jersey number of Tom						
3	Hansen	9		Function to be used = VLOOKUP						
4	Peter	12		Result to be put = E9						
5	Maxwell	66								
6	Baker	1								
7	Tom	5								
8	Chris	13								
9	Jake	35		Tom	=VLOOKUP(D9, A2:B9, 2, 0)					
10					VLOOKUP(lookup_value, table_array, col_index_num, [range_lookup])					

Fig. 5.127

Pressing enter would display 5.

In the case of 1 in the last argument, the table is to be organized in ascending order for Excel to be able to search through the player name.

Now, if we change the D9 cell from "Tom" into "Peter," we would obtain the jersey number for Peter.

	A	B	C	D	E	F	G
1	Player Name	Jersey Number					
2	Jesse	45		We want to determine the jersey number of Peter			
3	Hansen	9		Function to be used = VLOOKUP			
4	Peter	12		Result to be put = E9			
5	Maxwell	66					
6	Baker	1					
7	Tom	5					
8	Chris	13					
9	Jake	35		Peter		12	

Fig. 5.128

5.9   Conclusion

This chapter follows a practical approach and addresses some real-world problems that Excel users often stumble upon, such as getting rid of extra spaces within the text, detecting duplicate data, handling blank cells, and so on. Besides, the chapter also shows how to deal with equations and matrices in Excel. File management, forecasting unknown or future data, data manipulation techniques, etc., are neatly chalked out in this chapter. The application of Excel is immense in data science. Some basic and advanced functions of MS Excel that aid in the process of data analysis are also described in this chapter. Overall, this chapter will benefit all levels of users and will help nourish the most highly needed skills in working with MS Excel.

References

1. https://www.excel-easy.com/examples/forecast.html
2. https://trumpexcel.com/clean-data-in-excel/

Index

© The Author(s), under exclusive license to Springer Nature Switzerland AG 2021
E. Hossain, *Excel Crash Course for Engineers*,
https://doi.org/10.1007/978-3-030-71036-1

Printed in the United States
by Baker & Taylor Publisher Services